EARTH SCIENCES,
GEOGRAPHY
AND
CARTOGRAPHY

DE DIVERSIS ARTIBUS

COLLECTION DE TRAVAUX
DE L'ACADÉMIE INTERNATIONALE
D'HISTOIRE DES SCIENCES

COLLECTION OF STUDIES
FROM THE INTERNATIONAL ACADEMY
OF THE HISTORY OF SCIENCE

DIRECTION
EDITORS

EMMANUEL
POULLE

ROBERT
HALLEUX

TOME 53 (N.S. 16)

BREPOLS

PROCEEDINGS OF THE XXth INTERNATIONAL CONGRESS
OF HISTORY OF SCIENCE (Liège, 20-26 July 1997)

VOLUME X

EARTH SCIENCES, GEOGRAPHY AND CARTOGRAPHY

Edited by

Goulven LAURENT

BREPOLS

The XXth International Congress of History of Science was organized by the Belgian National Committee for Logic, History and Philosophy of Science with the support of :

ICSU
Ministère de la Politique scientifique
Académie Royale de Belgique
Koninklijke Academie van België
FNRS
FWO
Communauté française de Belgique
Région Wallonne
Service des Affaires culturelles de la Ville
  de Liège
Service de l'Enseignement de la Ville
  de Liège
Université de Liège
Comité Sluse asbl
Fédération du Tourisme de la Province
  de Liège
Collège Saint-Louis
Institut d'Enseignement supérieur
  "Les Rivageois"

Academic Press
Agora-Béranger
APRIL
Banque Nationale de Belgique
Carlson Wagonlit Travel -
  Incentive Travel House

Chambre de Commerce et d'Industrie
  de la Ville de Liège
Club liégeois des Exportateurs
Cockerill Sambre Group
Crédit Communal
Derouaux Ordina sprl
Disteel Cold s.a.
Etilux s.a.
Fabrimétal Liège - Luxembourg
Generale Bank n.v. -
  Générale de Banque s.a.
Interbrew
L'Espérance Commerciale
Maison de la Métallurgie et de l'Industrie
  de Liège
Office des Produits wallons
Peeters
Peket dè Houyeu
Petrofina
Rescolié
Sabena
SNCB
Société chimique Prayon Rupel
SPE Zone Sud
TEC Liège - Verviers
Vulcain Industries

D/2002/0095/14
ISBN 2-503-51360-3
Printed in the E.U. on acid-free paper

TABLE OF CONTENTS

INTRODUCTION

Goulven LAURENT

Les communications réunies dans ce volume ont ceci de commun qu'elles se rapportent à la Terre, à son " histoire naturelle ", à son exploitation et à sa description. On peut les classer en deux catégories distinctes.

La première partie comprend les communications qui ont trait à la manière dont l'homme a conçu ses rapports à la Terre et à la Vie sur la Terre. Nous avons d'abord la remarquable étude de Carlos Almaça, consacrée au naturaliste portugais Alexandre Rodrigues Ferreira et à son exploration du Brésil à la fin du XVIII⁵ siècle. Ensuite Martin Guntau retrace l'histoire de l'industrie minière depuis l'antiquité jusqu'à la veille de la révolution industrielle, en fournissant une bibliographie abondante, base précieuse pour des recherches plus approfondies. Andrès Galera Gomez montre ce que la paléontologie naissante du XVII⁵ siècle doit aux naturalistes napolitains Ferrante Imperato, Fabio Colonna et Agostino Scilla, et souligne combien leurs travaux ont eu d'influence sur le développement de cette science. Goulven Laurent traite de l'histoire récurrente d'un problème soulevé par les paléontologistes, depuis le début de la discipline, même par ceux d'entre eux qui sont évolutionnistes : celui de l'apparition brusque des espèces et des faunes, sans lien apparent avec celles qui les précèdent. Michael A. Cremo montre que l'histoire des sciences gagne à reprendre l'étude de certains cas qui ont été considérés trop vite comme résolus. Il prend comme exemple les découvertes de Boucher de Perthes à Moulin Quignon, suspectées à l'époque de fausseté pour des raisons qui n'ont rien à voir avec la science ! Un autre sujet d'études a été les tremblements de terre. Rebecca de Gortari et Maria Josefa Santos Corral ont exposé les conditions de la création, à la fin du XIX⁵ et au début du XX⁵ siècle, du service sismologique du Mexique, et l'importance de la collaboration internationale (même politique) à cette occasion. Svetlana Akhundova s'est attachée à décrire la manière dont on a tenté, depuis l'antiquité jusqu'à nos jours, de prédire les séismes. Elle expose les progrès réalisés de nos jours en URSS, et plus particulièrement dans son pays, l'Azerbaïdjan, depuis les catastrophes qui ont dévasté récemment ce pays.

La seconde partie est consacrée à la description de la Terre : la cartographie et la géographie. Pour continuer par l'Azerbaïdjan, nous prenons connaissance d'un article — malheureusement trop court ! — de Tamara Arkadiyevna Zolotovitskaya sur la nature de la radioactivité de son pays. Iaroslav Alexevitch Matviichine nous fournit un article très riche et très documenté sur les cartes de l'Ukraine réalisées par des voyageurs occidentaux du XIVe au XVIIIe siècle, qui nous fournit une mine de renseignements sur ce sujet. Alexei V. Postnikov expose les explorations réalisées par les voyageurs russes pendant plus d'un siècle — de 1741 à 1867 — dans la partie russe du continent américain, et souligne l'importance qu'elles ont présentée pour la connaissance, non seulement de ces territoires, mais pour celle de toute la région, et au-delà. L'histoire allemande se signale, elle aussi, par des avancées importantes. Ute Wardenga nous fait connaître les nouvelles directions de recherche dans lesquelles elle s'est engagée avec quelques collègues, et les débats d'idées auxquels elles ont donné lieu. D'autres innovations se font jour dans le domaine de la représentation de l'espace. Grigoriy Kostinskiy nous expose le changement dans les concepts et analyse en profondeur les différentes notions utilisées dans les travaux géographiques. D'une manière originale, et inattendue, Manfred Büttner montre l'influence que la Réforme protestante a exercée sur les études de géographie terrestre, en citant des textes de Melanchthon, Mercator et Keckermann, qui méritent en effet l'attention des historiens de cette discipline, auxquels Büttner fournit une abondante et précieuse bibliographie. Il n'y a pas que les concepts — si importants qu'ils soient — qui comptent. Steven L. Driever nous montre comment les représentations que l'on se fait d'un pays peuvent être amenées à se modifier. Il expose le cas significatif de la manière dont l'image de l'Espagne s'est transformée au début de ce siècle. Les scientifiques ne travaillent pas dans une tour d'ivoire. Marion Hercock étudie les processus qui ont conduit à des développements différents dans deux îles voisines en Australie, selon que les décideurs politiques se sont tenus en contact ou non avec des géographes. L'ensemble des communications présentées s'achève, dans ce recueil, comme elle avait commencé, par une étude sur une activité historique et géographique du Brésil. Alda Lucia Heizer nous entretient en effet des liens qui ont existé entre l'Institut historique de Paris et l'*Instituto Historico e Geografico Brasiliero*. Enfin, Marie-Claire Robic analyse les conditionnements politiques sous-jacents à la géographie de l'entre-deux guerres.

Au total, ce volume apporte un lot important d'idées neuves et de bases solides pour des études approfondies ultérieures.

Natural History and its Applications in the "Philosophical Expedition" of Alexandre Rodrigues Ferreira to Amazonia and the Mato Grosso (1783-1792)

Carlos ALMAÇA

INTRODUCTION

As trade with the Orient progressively declined throughout the 17th century, Portugal turned its attention back to Brazil. Here it hoped to find the tropical products to replace those of the Orient which were eagerly sought after in Europe : spices, metals and precious stones. Hence, the urgency behind the Marquês de Pombal's astute policy of pursuing the discovery and exploitation of the natural products that Brazil's immense territory might offer, and of introducing new technologies to improve their usefulness.

There was a shortage, however, of adequately trained scientists to whom the exploration and technological development of Portugal's overseas territories could be entrusted. For this reason, the Marquês de Pombal, following his fundamental reorganisation of the country's economy, set about restructuring the University in 1772 and engaged foreign lecturers for the course of Natural Philosophy. The responsibility for teaching and research in Chemistry and Natural History fell on the Italian Domingos Vandelli (1730-1816), born in Padua and deceased in Lisbon. His natural history collection was donated to the University (founded in 1290 in Lisbon and from 1537 definitively transferred to Coimbra), where it formed the embryo of the Museum of Natural History. An excellent organiser, Vandelli was the founder of Portuguese research into Natural History[1].

---

1. On Vandelli and his contributions to pedagogical and scientific organisation in Portugal see J.A. Simões de Carvalho, *Memória histórica da Faculdade de Filosofia*, Coimbra, 1872 ; W.J. Simon, *Scientific expeditions in the Portuguese overseas territories (1783-1808) and the role of Lisbon in the intellectual-scientific community of the late eighteenth century*, Lisboa, 1983.

It was at the reformed University that the naturalists received their theoretical training. Later, employed at the Real Museu e Jardim Botânico da Ajuda, founded in 1772 in Lisbon and also directed by Vandelli, they familiarised themselves with collections of natural products and the study of their applications during the years preceding the "philosophical expeditions". When it became possible to start the expeditions[2], in 1783, Alexandre Rodrigues Ferreira and two draughtsmen ("riscadores"), José Joaquim Freire and Joaquim José Codina, and a botanical gardener, Agostinho Joaquim do Cabo, were assigned to Brazil[3].

During this expedition of nearly 9 years, Ferreira and the team he directed produced around 100 manuscripts, 59 tables, 8 maps, 9 drawings, 9 black and white prints and 979 coloured prints of Amerindians, animals and views of cities, towns, villages, forts, buildings, rivers and waterfalls[4]. The greater part of these items were sent to Brazil in 1843 to be published and afterwards returned to Portugal. In fact, nothing came back.

The manuscripts signed by, or ascribed to, Ferreira, some with hundreds of pages, cover the extremely wide concept of Natural History of those days. I classify them under the following areas : history (9), geography (9), itineraries and logs of the expedition (8), hygiene (2), anthropology (11), ethnography (6), botany (8), agricultural sciences (5), plant technology (woods and plants for medicine, dyes and food) (8), zoology (13), remittance notes accompanying the collections (5), and miscellaneous (lectures, instructions, etc.) (11).

The doubles of some manuscripts and 210 original drawings remained, however, in Portugal and are now stored in the historical archive of Museu

2. As a result of the expeditions, the Real Museu stored important natural history and ethnographic collections, especially from Brazil. On the inventory of the collections housed in 1794 and appropriations perpetrated at the Real Museu see C. Almaça, *A natural history museum of the 18th century : the Royal Museum and botanical garden of Ajuda*, Lisboa, 1996.

3. On the A.R. Ferreira's expedition see W.J. Simon, *Scientific expeditions in the Portuguese overseas territories...*, op. cit. ; J.C. de Mello Carvalho, *Viagem filosófica pelas capitanias do Grão Pará, Rio Negro, Mato Grosso e Cuiabá (1783-1792)*, Belém, 1983 ; M.L. Rodrigues de Areia, "Perfil de um naturalista", *Memória da Amazónia* (1991), 13-75 ; A. Domingues, "As remessas das expedições científicas no norte brasileiro na segunda metade do século XVIII ", *Brasil* (1992), 87-93 ; C. Almaça, "Alexandre Rodrigues Ferreira e a exploração histórico-natural do Brasil ", *Oceanos*, 9 (1992), 54-57 ; C. Almaça, A. Domingues, M. Faria, *Viagem filosófica de Alexandre Rodrigues Ferreira*, Lisboa, 1992.

4. A "Catalogue " edited shortly after Ferreira's death, one copy of which is housed in the historical archive of Museu Bocage (ARF-23), mentions 171 manuscripts. Among them 61 are signed by Ferreira and four by Agostinho Joaquim do Cabo. The remaining 106 manuscripts are anonymous or by authors unrelated to the expedition. However, nearly 40 of the anonymous manuscripts were probably by Ferreira. See Costa e Sá, "Elogio ao Doutor Alexandre Rodrigues Ferreira ", *Memórias da Academia Real das Ciências de Lisboa*, 5 (1818), 58-89 ; A. Valle Cabral, "Notícia das obras manuscritas e inéditas relativas à viagem filosófica do Doutor Alexandre Rodrigues Ferreira ", *Anais da Biblioteca Nacional, Rio de Janeiro*, 1 (1876), 103-129, 222-247, 2 (1877), 192-198, 3 (1877), 54-67, 324-354 ; J.H. Rodrigues, "Alexandre Rodrigues Ferreira ; catálogo dos manuscritos e bibliografia ", *Anais da Biblioteca Nacional, Rio de Janeiro*, 72 (1952), 5-153.

Bocage (Lisbon). A few are remarkable for their interest and rigorous observation and have been selected for this paper.

DISEASES OF THE INDIANS

During his long experience in the Amazon jungle, Ferreira had the opportunity to observe the Indians and the way they were used and exploited by the white colonists. He expressed pity for the way they were treated and the lack of interest in their welfare. In a report he wrote on the " causes of diseases among the Indians " he suggests three main causes among the seven which he believed they fell victim to. Those causes were :

(a) " Unremitting and strenuous physical labour ". The Indians worked like veritable pack animals for white traders. They travelled along the rivers and whenever rapids were encountered they had to carry the boats and cargo overland on their back.

(b) " The nakedness of their bodies, permanently exposed to the vicissitudes of the weather ". On journeys lasting five, six or more months, the Indians took only a shirt, breeches and a sleeping mat. That was all they had to protect themselves against the sun, rain, cold and heat. If it was cold, they lit a fire and slept close to it.

(c) " The corruption of the food they eat and the impurity of the water they drink ". They ate salted meat and fish, poorly stored in the unventilated holds of the canoes. These provision rotted, causing serious illnesses.

Ferreira's concern for humane treatment and hygiene made considerable sense. Quite apart from the brutal working conditions imposed on them, the Indians were decimated by contact with diseases brought in by Europeans and Africans to which they were not immune[5].

THE RIO NEGRO : AN INEXHAUSTIBLE SOURCE OF NATURAL AND MAN-MADE PRODUCTS

To conclude his monumental *Body of general and particular History of the Rio Negro*, a work containing 13 reports, Ferreira summarised in the last report " everything on which he had written extensively ".

The Rio Negro, a tributary of the Amazon, in its 2.200 km long course has been discovered and navigated by the Portuguese since the 17th century. It boasts a huge number of islands and reaches 20 km in width. Around 30 other rivers flow into it, among them the Branco, also exploited by Ferreira. Its basin

---

5. On the uses to which the Indians were put and the culture clash with the whites, see L.F. de Alencastro, " A interacção europeia com as sociedades brasileiras entre os séculos XVI e XVIII ", *Brasil* (1992), 97-119 ; A. Domingues, " As sociedades e as culturas indígenas face à expansão territorial luso-brasileira na segunda metade do século XVIII ", *Brasil* (1992), 183-207.

was inhabited by numerous tribes, practically one per tributary, almost all of them practising cannibalism.

The banks of the river were extremely suitable for growing of indigo, coffee and tobacco. Indigo had started to flourish only a short time before, thanks to the factories that had been built along the upper and lower reaches of the Rio Negro. It was the best exported from Brazil to Lisbon, the amount of the export attaining 13 contos (13 million reis) in 1785 and 80 contos in 1787.

Coffee had not long been introduced into the Rio Negro District. It seems to have been brought from Africa to Surinam in 1718 and from Guiana to Brazil in 1720. Carefully cultivated it produced good yields, especially on the banks of Rio Negro. In the best years, the harvest of the District could attain 22 tons.

As for tobacco, after the sugar the most important export from Brazil until the end of the 18th century, only a little was grown along the Rio Negro. However, the total exported was very significant : 1.600 contos.

Ferreira thought there was a certain lack of drive about the agriculture of the Rio Negro and Pará Districts, that is, Amazonia. In his view the main causes of this state of affairs included :

(1) the laziness of the natives ;

(2) the shortage of labour because of the lack of black slaves, the excessive use of Indians for expeditions and the high death rate from epidemics ;

(3) the Europeans' contempt for work ;

(4) the ignorance of the correct agricultural techniques ;

(5) the hostility of the natives ;

(6) the malign effects of the " jungle products " trade ;

(7) damaging manufactures, etc.

The " jungle products " (" drogas do sertão ") such as, pitch (breu), wild and subspontaneous plants with medicinal properties, trees producing excellent wood for building, furniture or high-class joinery, were much valued by the export trade. Involving low investment and often benefiting from tax concessions or exemptions, their harvesting became an incentive to the colonisation of Amazonia and was encouraged from the second half of the 18th century[6]. However, it distracted the Indians for the best part of the year.

The manufactured products were just as varied : turtle oil for domestic lighting and to be eaten, pottery made in factories or by hand, fishing nets, straw hats, dyes, sugar cane extract, rum, etc. There were four potteries along the Rio Negro where jugs, pitchers, bowls, serving plates, chamber pots, dishes, stoves, lamps, kitchen pots, roofing tiles, bricks, candlesticks, and measuring jugs were made. The Indians made several domestic utensils by hand and other

6. A. Domingues, " Drogas do sertão ", in M.B. Nizza da Silva (coord.), *Dicionário da história da colonização portuguesa no Brasil*, Lisboa, 1994, 270-271.

goods like, for example, rum. This was made in small, animal-powered mills that crushed the sugar cane, named molinotes, which implied a low investment. In Ferreira's opinion these molinotes undermined agriculture because, given the simplicity and low cost of rum manufacture, the few workers there were preferred it to farming.

The local diet was based almost wholly on wild species, with fish and turtles playing a major role. The turtle, wrote Ferreira, " is the beef of Portuguese dining tables. It is eaten boiled, roasted, fried and stewed, as well as its eggs ". However, between its fishing and use for eating, the mortality was horrifying. From 1780 to 1785, 53.468 turtles were delivered to the District " corrals " and only 36.007 were used ; 17.461 have died before any use.

In times of famine, the Indians lived almost exclusively on manioc. From it, they prepared various foods and drinks :

(a) tapioca, or grated manioc ;

(b) carimáas, grated manioc rolled very fine ;

(c) " water flour ", that is, soaked manioc ;

(d) " dry flour ", not soaked manioc ;

(e) beijúz, cakes of manioc ;

(f) wine ;

(g) and spirits.

The Indians preferred fruit to meat or fish, using an extraordinary variety of plants. Ferreira refers to 120 plant species from which they ate the fruits, seeds or roots. They caught also a huge variety of animals for eating : mammals, birds, reptiles, fishes, ants, crabs and molluscs.

THE COMPREHENSIVE EXPLOITATION OF NATURAL PRODUCTS

One of Ferreira's constant concerns in describing whatever he encountered was its possible usefulness. This can be seen in the manuscript entitled " Report on the *Pirá úrucú* Fish ". Ferreira described the *Pirá úrucú* in Latin and in Portuguese, naming it *Paraensibus pira-urucu*. He never published, however, the description so that the valid name of the species, *Arapaima gigas*, is that given to it by Cuvier. Yet, the species had already been briefly described by another Portuguese naturalist before Ferreira, Frei Cristóvão de Lisboa (1583-1652)[7]. Its Indian name means " fish painted with úrucú ". Úrucú is the annatto or achiote, a starch extracted from the seeds of *Bixa orellana*, with which the Indians painted their bodies.

7. J. de Paiva Carvalho, " Comentários sôbre os peixes mencionados na obra *História dos animais e árvores do Maranhão por Frei Cristvóvão de Lisboa* ", *Arquivos da Estação de Biologia Marítima da Universidade de Ceará*, 4 (1964), 1-39 ; H. Nomura, *História da Zoologia no Brasil : século XVII*, Fundação Vingt-Un Rosado, 1996, 71-130.

The animal attains the greatest dimensions known in freshwater fish (up to 5 m long and 200 kg in weight). It was very abundant and reached large size, some specimens yielding around 50 Kg's of flesh[8]. The flesh was salted or dried for eating. Ferreira compared it, when dried, to the bacalhau (cod fish) eaten in Portugal. The tongue bone of the *Pirá úrucú* was used by the Indians as a grater for various seeds. The scales were the principal form of smoothing agent, being used by turners, carpenters and similar craftsmen.

THE " SHELL MINES " AND LIME PRODUCTION

The " shell mines " or " sernambi mines " are kjokkenmodings located on the coast, or the banks of rivers and lagoons near the shore. Essentially composed of shells of an edible marine lamellibranch, named sernambi (*Mesodesma mactroides*), these formations are the accumulation of cooking remains and skeletons of the humans who inhabited the area where they are found. The overwhelming presence of a marine mollusc indicates that the location of the " shell mines " are, or once were, close to the sea.

Ferreira described two " shell mines " so large that lime extraction had occurred over the last 80 years without any noticeable depletion. The mines contained " sernambi shells, petrified fish, human bones, shards of earthenware cooking pots and platters, bones of terrestrial animals, large conches, and the shells of oysters and many other shellfish ". In Ferreira's words " these remains provide physical evidence that the ocean once covered these hills, because these shellfish live only in the sea ". He concludes " this rarity is a prodigious source from which an immense fortune in lime can be extracted ".

DISCUSSION

The manuscripts summarised in the foregoing sections represent only a small part of the hundred or so that Alexandre Rodrigues Ferreira wrote during the nine years he spent exploring Amazonia and the Mato Grosso. They have been selected from those stored in the historical archive of the Museu Bocage because they reveal some of the more interesting aspects of Ferreira's scientific and humanistic development. They show, in fact, the great quality and thoroughness of the naturalist's observations, a remarkable attachment to the utilitarian view of the study of nature, and a humanitarian concern for the Indians that was far from common at his time or even today.

8. This fish continues to form the staple diet of Amazonian river dwellers in the present century because of the quality of its flesh. For this reason, and to prevent its extinction, a research programme was initiated in 1935 with a view of acclimatising it to other regions of Brazil. See J. de Paiva Carvalho, " Comentários sôbre os peixes mencionados na obra *História dos animais e árvores do Maranhão* por Frei Cristvóvão de Lisboa ", *op. cit.*

Ferreira was a meticulous observer, Cartesian in his methodology and taxonomic in his presentation of results. It is extremely interesting to note the way he reduces the causes of the Indians' diseases, the indifferent agricultural performance in Rio Negro or the ways of using manioc to a small number of categories, emphasising the principal ones where appropriate.

He encountered insuperable difficulties, however, with the nomenclature and taxonomy of the wide biological diversity that he observed. The great majority of neotropical species had not yet been described or scientifically named, at least in the bibliography he took with him on the expedition[9]. For this reason, Ferreira had to transliterate the local names and create ecological classifications (" lacustrine " or " riverine " fishes), practical ones (" cultivated " or " wild " native fruits, wild plants from which they " made wine ", or of which they " ate the seeds " or " consumed the roots ") or mixed classifications (Linnean and ethnobiological ones).

The last group is particularly interesting. In listing the wide variety of animals that the Indians hunted or fished, Ferreira uses Linnean classes and orders (*Mammalia : Primates, Bruta, Ferae, Glires, Pecora, Belluae ; Aves : Pica, Anseres, Grallae*, etc.). However, within each order, he resorts ethnobiological nomenclature which allows him, by means of a system of primary and secondary names[10], to identify the animals caught by the Indians. This was because, as has already been remarked, Ferreira did not have access to the bibliography indispensable to the identification of Linnean species, particularly the 10th (1758) and later editions of *Systema naturae*, which contained copious information on neotropical fauna.

The way he creates the names is very curious. The primary names seem always to be formed by transliterating native names. For example, sloths (*Bradypus, Choloepus*) are called *ay,* manatees (*Trichechus*) *yuarauá,* marmosets (*Callitrix, Marikina, Cebuella*), *xaguim,* deer (*Odocoilus, Mazama*) *suassú,* peccaries (*Tayassu*) *tayaçu,* etc.

With the secondary names, the super ordinate category is always represented by a native name, but this is not always the case with the included constituent (subordinate category) ; sometimes it is a Portuguese word. For example : he uses in both categories (super ordinate and subordinate) Indian names when he distinguishes three taxa in sloths — *ay guaçu, ay merim* and *ay tatá* —, two in marmosets — *xaguim tinga* and *xaguim poxuna* —, or four in deer — *suassú apara* (*Odocoilus virginianus*), *suassú tinga* (*Mazama sim-*

---

9. This bibliography is known from a manuscript that formed part of the ARF archive in Museu Bocage, which caught fire on the 18th March 1978. Fortunately, it had been copied prior to the fire and reproduced by W.J. Simon, *Scientific expeditions in the Portuguese overseas territories..., op. cit.* The natural history bibliography comprised *Flora Guyana,* by Aublet, *Historia naturalis Brasiliae,* by Marcgrave and Piso, *Systema naturae,* 1755, *Genera Plantarum* and *Species Plantarum,* by Linnaeus, *Histoire des Poissons,* by Golvan, and *Histoire des Insectes.*

10. See B. Berlin, *Ethnobiological classification,* New Jersey, 1992.

plicicornis), *suassú anhanga* and *suassú caapora*. But, in manatees, he uses the Indian name to the super ordinate category and a Portuguese name to the subordinate : *yuarauá* ordinário (*Trichechus manatus*) and *yuarauá* de manteiga (*Trichechus inunguis*). In peccaries Ferreira applies Indian names to both categories, distinguishing the varieties by Portuguese names : *tayaçu uaia, tayaçu tastitu* and *tayaçu caapora*, the last with the varieties " com queixada branca " e " sem queixada ".

The utilitarian view of the understanding of nature is very typical of Ferreira, as with other, prior Portuguese explorers in Brazil[11]. This view is prominent in the manuscripts summarised in this paper, though as has been pointed out[12], the study of natural products was perhaps not the most important objective of Ferreira's expedition. However that might be the identification of species, descriptions of geological formations, the uses which could be made of them, the technologies used to exploit them and improvements that might be introduced are constant concerns in the manuscripts of the expeditions. Whether these were the result of the influences of the period, of the lessons of his master Vandelli or of his own way of looking the world, the fact is that Ferreira evidenced these concerns very early on.

Thus, in a paper he presented at the Academy of Sciences of Lisbon on 21st November 1781 (he was then 25 years old), entitled " The abuse of conchology in Lisbon ; to serve as an introduction to my theology of worms ", Ferreira vehemently criticises those who interest themselves in the study of nature without concern for the country's economy. This attitude, further reinforced by an anti-conservation mind which is surprising to the modern reader, was typical of Vandelli's teaching. In the book he wrote for Portuguese students[13], Vandelli asserts that the study of Zoology does not consist " ...of a simple knowledge of the name of each animal ; rather, it is necessary to know as much as possible about its anatomy, its way of living, breeding, its food, the uses that may be made of it ; and to know how to preserve those that are essential to the economy, attempting to discover the uses they have which we do not at present know, and eliminate those that are harmful, or defend oneself from them ".

Ferreira followed these instructions to the letter. His splendid monograph on the Rio Negro is a compendium of natural products, their applications and

11. See C. França, " Os portugueses do século XVI e a História Natural do Brasil ", *Revista de História*, 15 (1926), 35-166 ; C. Almaça, " Os portugueses e o conhecimento das faunas exóticas ", *Oceanos*, 6 (1991), 52-63 ; " Os portugueses do Brasil e a Zoologia pré-lineana ", *A Universidade e os Descobrimentos* (1993), 177-193.

12. A. Domingues, " Um novo conceito da ciência ao serviço da *razão de estado* : a viagem de Alexandre Rodrigues Ferreira ao norte brasileiro ", in C. Almaça, A. Domingues, M. Faria (eds), *Viagem filosófica de Alexandre Rodrigues Ferreira*, Lisboa, 1992, 17-32.

13. *Diccionario dos termos technicos de Historia Natural extrhidos das obras de Linnéo, com a sua explicação, e estampas abertas em cobre, para facilitar a intelligencia dos mesmos e a memoria sobre a utilidade dos Jardins Botanicos que offerece à raynha D. Maria I, nossa Senhora, Domingos Vandelli*, Coimbra, 1788.

value as exports. When he laments the damage that had been done to the turtle population, it was not their extinction that concerned him ; perhaps he was not even aware of the possibility. Rather, it was their unit cost and the denial of the ones that died as food for the inhabitants. When he writes about the " shell mines ", those magnificent natural monuments steeped in history, he does not exhibit any concern for their preservation, appearing to content himself with the comforting certainty that they will yield a great deal of lime. Finally, when he describes the *pirá-úrucú*, which he judged, rightly, to be a species new to science, it is not this fact that raises its value in his eyes, but rather its size and the quantity of food that it signifies, and the way the entire fish can be made use of.

Finally, the humanitarian side of the naturalist. Ferreira did not hesitate to contrast the conditions of life in the Amazon jungle enjoyed by the whites and their Indian servants. The brutalities to which the Indians were subjected are uncompromisingly documented. It should be noted that Ferreira's manuscripts were intended to be read by government and local administrative bodies.

Ferreira's magnificent expedition did not have enduring results. The " Natural History of Pará ", which was in the course of preparation in 1804, with numerous copper engravings[14], was brought to an untimely end by the expropriations of General Lannes (1804) and Etienne Geoffroy Saint-Hilaire (1808)[15]. Thus ended an interesting, but brief, phase of the Portuguese Enlightenment.

14. Letter of 9th March 1804 in which the Portuguese Government commissioned an engraver to work under Ferreira's direction on the preparation of the illustrations for " História Natural do Pará ", *Historical archive*, MB, CN/V-43. He engraved an appreciable number of copper plates with Brazilian themes, descriptions of which are contained in manuscript ARF-23 (see fn. 4).

15. See C. Almaça, *A natural history museum of the 18th century : the Royal Museum and botanical garden of Ajuda, op. cit.*

GEOLOGICAL AND MINERALOGICAL KNOWLEDGE AND THE FIELD OF MINING BEFORE THE INDUSTRIAL REVOLUTION

Martin GUNTAU

The development of geological and mineralogical knowledge in the history of mankind is due to various reasons and different contexts. Obviously, practical needs of people, on the one hand, constituted one of the first important sources of experience and knowledge about the nature of the earth, which most of all were to mining. But, on the other hand, and at the same time medical ideas promoted the knowledge of minerals decisively. Thirdly, a very immediate kind of people's curiosity acted as propelling force to learn about the phenomena of their natural surrounding (like colorful minerals, metals, rocks, caves in the earth or earthquakes), the curiosity for geological knowledge. Forthly, — last but not least — in history, since Graeco-Roman times the search of knowledge about the earth has been stimulated by philosophical maxims and religious belief.

It goes without question that contexts of geological and mineralogical knowledge and the field of mining have always been of extraordinarily great significance. Although mining reaches back into the beginning of human history, the first knowledge of the nature of the earth did by no means emerge as geology or mineralogy in terms of scientific disciplines.

1. Miners of ancient times and the middle ages had knowledge of — and experience with the nature of the earth. These immaterial elements of mining remained an empirical component and yet only an integral part of mineral production without developing into a separate subject area. The reasons for this lay in the immature knowledge of minerals and geological phenomena and their meaning for the field of mining. Mining and miners had a very low social status at that time and were, therefore, largely ignored by the educated. Mining was often carried out under hard natural conditions, mostly at big distances from the ancient towns and cultural centres, by producers who were mainly

slaves, prisoners of war or condemned convicts[1].

Of course, we can nevertheless find various examples to illustrate how, even in ancient and medieval times, observations from mining together with other kind of knowledge were documented in a written form. We should draw the attention to the map from about 1300 B.C., a map of the Wadi Hammamat in Egypt, with a description of the " ways that lead to gold " (Turin papyrus), which was transmitted as a testimony for tremendous mining in the Pharaohs' Empires[2]. This papyrus can be considered as a mining map under the conditions of that time. In 355 B.C., the Greek Xenophon, a student of Socrates, provided — together with other pertinent explanations — the first geological descriptions of the Laurion ore deposit, where the Athenians exploited lead and silver. A number of mineralogical findings is documented in the mineralogical texts of Theophrastos (374-288 B.C.) or in those of the Roman natural historian Plinius Secundus (23-79)[3].

Similarly, knowledge of miners can be traced back to the Middle Ages in geological and mineralogical scripts of the Moslems Al Biruni (973-1048) and Ibn Sina (980-1038) and in Christian Central Europe in " liber lapidum " by Marbode (1035-1123) and in Albertus Magnus (1193-1280) " de lapidibus "[4].

All of them, though, did not constitute systematic surveys of geological and mineralogical knowledge about mining, because they were drawn up under medical, mythical or more general points of view. But mining had always been a significant source of knowledge without itself dependent on these elements of knowledge, because empirical aspects in mining as such had considerably proved sufficient.

2. A far-reaching change in the relations between mining and geological knowledge had been taken place in Europe since the beginning of the 16th century in the context of the Renaissance. Educated people changed their attitudes towards nature fundamentally. During the Renaissance scholars and artists (" artefici ") discovered mining — a " high-tech production " at that time — and this was the reason for developing systematic descriptions of the nature of the earth from the experiences of the miners. Starting in 1500, authors, who had partly studied at German or Italian universities, wrote mining books — mostly in form of manuscripts, which contained information about metals, minerals, ore veins, rocks and mountains.

    1. A. Eggebrecht, " Die frühen Hochkulturen : Das alte Ägypten ", in H. Schneider (ed.), Geschichte der Arbeit, Frankfurt/M, 1983, 67-70.

    2. H. Wilsdorf, Bergleute und Hüttenmänner im Altertum, Freiberger Forschungshefte D 1, Berlin, 1952, 28-47.

    3. Cf. R. Halleux, Le problème des métaux dans la science antique, Paris, 1979, 115-128 (Bibliothèque de la Faculté de Philosophie et Lettres de l'Université de Liège, Fasc. CCIX).

    4. Cf. K. Mieleitner, " Geschichte der Mineralogie im Altertum und im Mittelalter ", Fortschritte der Mineralogie, 7, (1922) 454-480.

On this background, the famous works of Vanuccio Biringuccio (1480-1538), Georgius Agricola (1494-1555) or Lazarius Ercker (1530-1594) were brought into existence. Although mining and metallurgy had been practised for millennia only then through the connection of education and mining production the chances were opened for describing technological processes and their natural conditions systematically. Agricola wrote together with his geological works[5] a survey about the then known ore deposits[6], a systematic and detailed summary on minerals[7] and , especially, fundamental insights into the natural history of ore deposits[8] in his main work *De re metallica*.

These works are not only typical for a new and high quality of scientific character. They are systematically structured and the statements offered can be verified. This knowledge was no doubt very useful for mining in those times in one or another ; nevertheless, ore and mineral production was still possible without a systematic acquisition of that knowledge. Similarly, an academic training of technicians and civil servants for mining had not developed as a necessary demand either.

In the second half of the 18th century a quantitatively large production growth in mining could be observed alongside with a significant increase of manufactories and at the beginning of introducing machinery. The rising production output in many countries of this time provoked an increasing interest in new deposits and a more effective exploitation of the new resources. Additionally, there was a crisis in mining in some of the countries.

In Saxony, mining which had been booming was ruined by the seven year war (1756-1763) so that needed to be reconstructed[9]. There were also big problems with the mercury production in Spain[10]. This was why the production of gold and silver sank dramatically. Similarly, the very England-oriented mining export of iron and copper in Russia had lost effectivity, which made reforms necessary[11]. Especially complicated was the situation, where mining had been performed intensively for centuries and stocks run short.

As happened in the age of enlightenment the potencies of science were asked to come into remembrance. Geological and mineralogical knowledge

5. G. Agricola, " De ortu et causis subterraneorum " [1544], *Agricola-Werke*, vol. III, Berlin, 1956, 45-211.

6. G. Agricola, " De veteribus et novis metallis " [1546], *Agricola-Werke*, vol. VI, Berlin, 1961, 57-138.

7. G. Agricola, " De natura fossilium " [1546], *Agricola-Werke*, vol. IV, Berlin, 1961.

8. G. Agricola, " De re metallica " [1556], *Agricola-Werke*, vol. VIII, Berlin, 1974, 53-131.

9. H. Baumgärtel, *Bergbau und Absolutismus. Der Sächsische Bergbau in der zweiten Hälfte des 18.Jahrhunderts und Maßnahmen zu seiner Verbesserung nach dem Siebenjährigen Krieg*, (Freiberger Forschungsheft D 46), Berlin, 1963, 63-91, 134-139.

10. *Cf.* M. Guntau, " Die Genesis der Geologie als Wissenschaft ", *Schriftenreihe f. Geol. Wiss.*, H. 22 (1984), 57-58.

11. *Cf.* M. Guntau, " Die Genesis der Geologie als Wissenschaft ", *Schriftenreihe f. Geol. Wiss.*, H. 22 (1984), 54-57.

had become independent of mining and stabilized since the Renaissance to a growing degree without losing contact to its sources. This was shown by the works of civil servants or scholars active in the field of mining and geology like :

Johann Friedr. Wilh. Charpentier[12]	1728-1805	Saxony
Abraham Gottlob Werner[13]	1749-1817	Saxony
Johann Gottlob Lehmann[14]	1719-1767	Prussia
Carl Abraham Gerhard[15]	1738-1832	Prussia
Georg Christian Füchsel[16]	1722-1773	Thuringia
John Strachey[17]	1671-1743	Britain
John Williams[18]	1730-1797	Britain
Déodat Gratet de Dolomieu[19]	1750-1801	France
Axel von Cronstedt[20]	1722-1765	Sweden
Johann Jakob Ferber[21]	1743-1790	Sweden
Antonio Vallisnieri[22]	1661-1730	Italy
Giovanni Arduino[23]	1714-1795	Italy
Michail Wassil. Lomonossow[24]	1711-1765	Russia
Christoph Traugott Delius[25]	1728-1779	Austria

12. J.F.W. von Charpentier, *Mineralogische Geographie der chursächsischen Lande*, Leipzig 1778.

13. A.G. Werner, *New theory of the formation of veins ; with its application to the art of working mines* [1791], Edinburgh, 1809.

14. J.G. Lehmann, *Versuch einer Geschichte von Flötz-Gebirgen, betreffend deren Entstehung, Lage, darinnen befindliche Metalle, Mineralien und Fossilien*, Berlin, 1756.

15. C.A. Gerhard, *Versuch einer Geschichte des Mineralreichs*, Berlin, 1781-1782 (2 vols).

16. G.Ch. Füchsel, *Historia terrae et maris ex historia thuringiae, per montium descriptionem, Actorum Academiae electoralis moguntinae scientiarum utilium quae Erfordiae est*, Erfordiae, II, 1761, 44-254.

17. *Cf.* R. Porter, *The Making of Geology*, Cambridge, 1977, 119.

18. J. Williams, *The natural history of the mineral kingdom*, Edinburgh, 1789 (2 vols).

19. *Cf.* G. Schmeisser, *Beyträge zur näheren Kenntniß des gegenwärtigen Zustandes der Wissenschaften in Frankreich*, Hamburg, 1797, 107.

20. A. von Cronstedt, *Mineralgeschichte des Westmanländischen und Delekarlischen Erzgebirges*, Nürnberg, 1781.

21. J.J. Ferber, *Beyträge zu der Mineralgeschichte Böhmens*, Berlin, 1774 ; *Versuch einer Oryktographie von Derbyshire in England*, Mietau, 1776 ; *Lettres sur la minéralogie et sur divers autres objets d'histoire naturelle de l'Italie...*, Strasbourg, 1776 ; *Neue Beiträge zur Mineralgeschichte verschiedener Länder*, Mietau, 1778 ; *Physikalisch-Metallurgische Abhandlung über die Gebirge und Bergwerke in Ungarn...*, Berlin/Stettin, 1780 ; *Mineralogische und metallurgische Bemerkungen, in Neuchatel, Franche-Comté und Bourgogne im Jahre 1788 angestellt*, Berlin, 1789.

22. *Cf.* E. Vaccari, " Mining and knowledge of the earth in the 18th century Italy " (paper of the XXth ICHS, Liège, 1997, in print).

23. *Cf.* E. Vaccari, *Giovanni Arduino (1714-1795)*, Firenze, 1993.

24. M.W. Lomonossow, *Erste Grundlagen der Metallurgie oder des Hüttenwesens*. Zweite Beilage : *Über die Erdschichten* [1761], vol. I, Ausgewählte Schriften, Berlin, 1961, 435-549.

25. Chr.T. Delius, *Abhandlung von dem Ursprung der Gebirge und der darinnen befindlichen Erzadern oder der sogenannten Gänge und Klüfte, ingleichen von der Vererzung der Metalle*, Leipzig, 1770.

Johann Ehrenreich Fichtel[26]   1732-1795   Austria

Ignaz von Born[27]   1742-1791   Austria

On the other hand, awareness for the significance of geological knowledge for mining was documented in a number of demands as for example by C.F. Zimmermann in 1746, according to which every " sovereign should perform a general examination of his ore mountains "[28]. Together with the need of geological and mineralogical knowledge in mining itself maps of geological making became relevant. And besides, since 1765, mining academies in addition to universities were founded in many countries[29] to provide education and training for civil servants with special technological and geological knowledge to qualify them for solving problems as were mentioned before. But geology did not serve as a kind of mining's maid only. But it was stimulated in its own development by important impulses and opportunities.

If we try and overlook the relation between mining and geological knowledge in these big periods of time, we are offered the opportunity of a certain historical order.

A. In ancient and medieval times material and immaterial aspects of the production processes in mining were closely interwoven that only little experience and knowledge from this field could enrich the knowledge about the nature of the earth. Educated people described mineralogical and geological knowledge of miners in their works sporadically and non-systematically only.

B. In the context of the Renaissance technological and scientific questions of mining and metallurgy had been made subject matter of a specific literature since the beginning of the 16th century. Scholars, esp. physicians and lawyers, developed a special interest in the production of precious and non-ferrous metals, glass, salt, colours and other substances. There were publications of scientifically ambitious works on the classification and systematic description of minerals, on geological phenomena in nature, methods for the search of ore deposits, the origin and structure of mountains, volcanoes and earthquakes. Those descriptions were relevant for mining, not necessarily indispensable for production yet.

C. At the time of transition from manufactories to the use of machinery (in the 18th century) the requirement for several raw material resources was increasing, which could not be met satisfactorily in all countries. Geological and mineralogical knowledge was gaining respect and value with the intensive

26. J.E. Fichtel, *Nachrichten von den Versteinerungen und Mineralien Siebenbürgens*, Nürnberg, 1780.

27. I. von Born, *Briefe über mineralogische Gegenstände auf seiner Reise durch das Temeswarer Bannat, Siebenbürgen, Ober- und Nieder-Hungarn*, Frankfurt, Leipzig, 1774.

28. C.F. Zimmermann, *Ober-Sächsische Berg-Academie...*, Dresden, Leipzig, 1746, 134.

29. M. Guntau, " Geologische Institutionen und staatliche Initiativen in der Geschichte ", in M. Büttner, E. Kohler (eds), *Geosciences/Geowissenschaften, Proceedings of the Symposium of the XVIIIth ICHS at Hamburg-Munich 1989*, part III, Bochum, 1991, 229-240.

use of existing deposits and the search of new ones as well as with the beginnings of geological mapping. Consequently, in the 18th century an extensive geological and mineralogical literature came into being, summarising all the corresponding knowledge and providing it for mining and usage.

And — last but not least — the requirements of mining were responsible too for the development of mineralogy and geology as independent natural sciences around 1800.

NÁPOLES Y LOS ORÍGENES DE LA PALEONTOLOGÍA

Andrés GALERA

GLOSSOPETRAE, TIEMPO DE FÓSILES

Glossopetrae es la denominación latina y genérica acopiada para las piedras de forma longitudinal muy abundantes en la isla de Malta. Por su estilización la tradición las reconoce como lenguas pétreas de serpiente, las considera un accidente de la naturaleza en la génesis de los seres vivos, atribuyéndolas propiedades farmacológicas contra la mordedura de los ofidios[1]. En realidad los pretendidos artefactos minerales son dientes de escualo fosilizados. El descubrimiento de esta verdad nos conduce al Nápoles de Ferrante Imperato y Fabio Colonna, estamos en los inicios de la paleontología[2].

Si en el discurso preliminar de su obra *Recherches sur les ossemens fossiles de quadrupèdes* Cuvier reconoce su cualidad de anticuario de la vida, un saber necesario para descifrar la historia del globo y reconstruir mediante el testimonio fósil los extintos seres que habitaron en el pasado, de suerte que hay un ayer, un hoy y un mañana de las especies que pueblan la Tierra, para los naturalistas del siglo XVII los restos fósiles son un problema de forma y materia, lejos de la dimensión temporal que tendrán una vez aceptado su valor orgánico.

El debate paleontológico propio del seicento napolitano tiene sentido en la definición de un espacio y en la elección de un método científico. La institución es el museo de Historia Natural regentado por Ferrante Imperato, y el objeto de estudio las piezas que componen una notable colección de muestras fósiles : "Pero todo aquél que viera, no sólo lo que yo tengo, sino lo que

---

1. F.M. Pompée Colonna, *Histoire naturelle de l'univers*, II, Paris, 1734, 301 (4 vols), recoge esta tradición. Michele Mercati (1541-1593) en su colección de Minerales y fósiles formada en el Museo Vaticano, Metallotheca Vaticana, incluía las glossopiedras entre los metales ; *cf.* M. Mercati, *Metallotheca. Opus posthumum*, Roma, 1717, 333-334.

2. Sobre el tema véase los trabajos de Nicoletta Morello, *La nascita della paleontologia nel seicento. Colonna, Stenone e Scilla*, Milan, 1979, 64-69 y 148-151 ; y "De Glossopetris dissertatio : the demonstration by Fabio Colonna of the true nature of fossils ", *Archives Internationales d'Histoire des Sciences*, 31 (1981), 63-71.

guarda Imperato en su Museo, todos aquellos géneros de caparazones convertidos en piedras o rellenos de una variedad de concreciones de piedra, no dudamos que deje de aprobar por experiencia la opinión de los antiguos " recuerda Fabio Colonna en su *De Glossopetris dissertatio*[3], partitura donde suenan los acordes de la futura paleontología. Si con su libro *Historia Naturale* Imperato realiza aportaciones novedosas, pues en sus páginas se reconocen como restos orgánicos no sólo las habituales incrustaciones conchíferas provinientes de las zonas montañosas, " vestigios de las inundaciones marinas ", sino que se toma partido favorable en casos más complejos y de mayor escepticismo como el Megaladon, conocido en la época como Bucardia y caracterizado por su semejanza al corazón del buey, y se identifica la primera rudista fósil, que denomina *corno di Ammone*[4], con su dissertatio Colonna se anticipa a Nicolás Sténon, al *Champollion* de la geología, como le ha bautizado Gabriel Gohau por establecer las reglas para la lectura de los estratos geológicos[5]. Cuando en 1666 Sténon estudia la anatomía del tiburón y comprende la relación entre las glossopetrae y los dientes de escualo[6], habían transcurrido cinco décadas desde que Colonna dio a conocer su significado fósil y más de un siglo desde que Leonardo da Vinci relacionase las imaginarias " lenguas de serpiente " con restos fósiles correspondientes a dientes de peces[7]. El hecho es algo más que una mera cuestión de anticipación y autoría, trasciende al ámbito ideológico. *De Glossopetris dissertatio* toma forma como una reflexión sobre el ser vivo acorde con una interpretación perfeccionista de la vida, pues la " naturaleza no hace nada en vano ", como reza el lema clásico[8]. Al aplicar este principio a la polémica paleontológica sobre el origen de los fósiles — orgánico o pétreo —

3. F. Colonna, *De glossopetris dissertatio*, Romae, 1616. Citamos por la traducción castellana incluida en J.M. Artola, A. Galera, " El tiempo biológico ", *Asclepio*, 46/2 (1994), 228.

4. F. Imperato, *Dell'Historia naturale*, Napoli, 1599, 665-667. Sobre el tema véase C.F. Parona," Saggio bibliografico sulle rudiste ", *Boll. R. Comit. Geol. d'Italia*, 6/1 (1916), 1-78 ; B. Accordi, " Ferrante Imperato (Napoli, 1550-1625) e il suo contributo alla storia della geologia ", *Geologica Romana*, 20 (1981), 43-56. Una breve relación de los fósiles custodiados en el museo de su padre se reproduce en la obra de Francesco Imperato, *Discorsi intorno a diverse cose naturali*, Napoli, 1628.

5. G. Gohau, *Une histoire de la géologie*, Paris, 1990, 75. *Cf.* también G. Gohau, *Les sciences de la terre aux XVIIe et XVIIIe siècles : naissance de la géologie*, Paris, 1990. Los principios geológicos de Sténon se encuentra recogidos en su obra *De solido intra solidum naturaliter contento Dissertationis Prodromus*, Florencia, 1669. Stephen Jay Gould en su obra *Hen's Teeth and Horse's Toes : Further Reflections in Natural History*, Penguin, 1984, realiza un certero análisis de la contribución de Sténon a la geología.

6. En F. Redi, *Osservazioni intorno agli animali viventi che si trovano negli animali viventi*, se relata la disección de Sténon ; citamos por la edición incluida en *Opere di Francesco Redi*, Florencia, 1858, 301. *Cf.* G. Gohau, *Une histoire de la géologie, op. cit.*, 69 y 72. Sténon conocía los trabajos de Colonna, como lo demuestra la referencia que hace de él en el volumen 2° de las *Actas Danesas* publicadas en 1666 (*cf.* F. Redi, *ibidem*, 401).

7. L. da Vinci, *Escritos literarios y filosóficos*, Madrid, 1930, 95.

8. J.M. Artola, A. Galera, *El tiempo biológico*, 223. Esta máxima general, por ejemplo, la repite Robert Hooke en 1665 en su *Micrografía* también para establecer el origen orgánico de los fósiles, rechazando la supuesta " virtud plástica " de la Tierra necesaria para la creación expontánea de estos objetos. Robert Hooke, *Micrografía*, Madrid, 1989, 339-343.

a su catalogación como animales o como unidades minerales, Colonna concluye la condición orgánica de las muestras ante la imposibilidad de que partes de un ser vivo puedan generarse con independencia del organismo al que pertenecen ; sería el fin de un sistema perfecto. " Hubieran sido elaboradas en vano por la naturaleza si no fuesen verdaderos dientes, y no piedras, parte de un animal muerto "[9] ; las partes son instrumentos y como tales necesitan del conjunto, del ser vivo, aisladas carecen de identidad. El fenómeno biológico no participa del azar, la génesis de las especies responde a un diseño perfeccionista donde el conjunto identifica a los elementos excluyendo cualquier individualidad, que cuando se manifiesta es fruto de la degradación corporal. Este ideario filosófico tiene su componente empírico, es una teoría apoyada en hechos. Determinar el origen orgánico de un fósil, rechazar su vínculo lapídeo, viene avalado por la forma y también por la materia : " Nosotros decimos que este tipo de formaciones no es de piedra apoyándonos en el aspecto, en la figura y en la sustancia que la compone "[10]. Eliminado el azar, aceptado el simbolismo biológico, la correspondencia formal de la muestra con especies vivas comfirma la hipótesis — un principio analógico que Sténon repetirá en su Prodromus al afirmar que " los cuerpos que asemejan plantas y animales encontrados en la tierra, tienen el mismo origen que las plantas y los animales que ellos representan "[11]. Además, cuando el fósil es óseo y no un modelo petrificado, como sucede con las glosopiedras, el análisis químico es una prueba irrefutable. La formación de carbón procedente de la combustión del hueso revela su naturaleza animal : " Ahora bien, dado que estos dientes cuando se queman se reducen rápidamente a carbón y la toba unida a ellos no, resulta claro que son de hueso y no de piedra "[12]. Para el geólogo británico Lyell, el sabio napolitano llegó aún más lejos en la ciencia paleontológica, pues supo diferenciar los tres estados de conservación de los restos fósiles — vaciado, molde y materia orgánica —, y tuvo " el mérito de ser el primero en observar que entre las conchas fósiles unas pertenecen a testáceos marinos y otras a testáceos terrestres "[13].

El debate sobre las glosopiedras tuvo continuidad. En 1670 el pintor y diletante naturalista siciliano Agostino Scilla publica en Nápoles *La vana speculazione disingannata dal senso*. La obra participa y desarrolla el argumento orgánico expuesto por Colonna y Sténon : " Se muy bien que los corales, las conchas, los dientes de lamia y de canis, y los erizos, etc., son verdaderos corales, conchas, dientes, caparazones y huesos petrificados, y no de piedra formados "[14]. Su tesis repite el modelo precedente, y la unidad genésica de los

9. J.M. Artola, A. Galera, *El tiempo biológico, op. cit.*, 225.

10. *Ibidem*, 222.

11. G. Gohau, *Une histoire de la géologie, op. cit.*, 73.

12. J.M. Artola, A. Galera, *El tiempo biológico, op. cit.*, 222.

13. Ch. Lyell, *Principles of Geology*, I, Londres, 1867-8,35 (2 vols).

14. A. Scilla, *La vana speculazione disingannata dal senso. Lettera responsiva circa i corpi marini, che petrificati si trovano in varij luoghi terrestri*, Napoli, 1670, 129.

seres vivos aparece como la principal cuestión a resolver para determinar la condición orgánica de los fósiles que, ahora, además, se analiza bajo el prisma de la reproducción. El desarrollo embrionario es el novedoso razonamiento que sustenta la hipótesis del origen orgánico de los fósiles : " primero se formaría el huevo y luego el animal, o bien un embrión completo del pequeño animalito y no una porción del mismo "[15]. En definitiva, la respuesta de Scilla expresa una consideración metodológica. El empirismo biológico sustituye a la especulación filosófica, " a las sombrías astracciones de los metafísicos ", que no es el camino elegido por él para descubrir la verdad sobre los cuerpos naturales[16]. Y la cuestión a debate no es otra que interpretar la vida bien como el fenómeno determinista propio de una naturaleza ordenada bien como el principio casual, azar, que conlleva admitir la condición pétrea de los fósiles como resultado de la acción indiscriminada de la naturaleza. Con su teoría embriológica Scilla niega la casualidad como ley natural. Su hipótesis se acerca a un ideario epigenésico afín al lema ex ovo omnia que rige la obra de su contemporáneo W. Harvey[17], en sintonía con el rechazo que, indirectamente, se realiza de la generación espontánea en las primeras páginas de la speculazione a través de la obra de Francesco Redi Esperience intorno alla generazione degl'insette[18]. Definir la vida terrestre como una realidad fisiológica lleva implícito interesantes connotaciones conceptuales. La consideración de vivo e inerte, la contraposición de animal y vegetal frente a mineral, no es ya una simple cuestión de forma, de cualidades anatómicas, sino la consideración del individuo como el resultado de un proceso fisiológico que determina su tipología. La morfología no es causa sino consecuencia. El ser vivo alcanza un nivel conoscible superior, que camina hacia esa visión integradora de la especie y su hábitat.

La consecuencia inmediata de aceptar la condición orgánica de los fósiles es la necesidad de considerar el mar como el medio de transporte de las diferentes especies acuáticas que representan, por las distintas y distantes regiones que habitaron en el pasado. La distribucción geográfica fue objeto de polémica. Para Colonna y Scilla — uno y otro plantean el problema del origen animal del registro fósil con independencia de su procedencia —, la fosilización es un fenómeno continuo en el tiempo, acorde con una historia geológica de la tierra caracterizada por las inundaciones : " Esto no sólo aconteció en tiempos del diluvio universal, sino que en otros siglos y en otros sitios el mar y la tierra cambiaron alternándose entre sí ", afirmaba Fabio Colonna distanciándose del precepto bíblico[19]. Naturalistas más respetuosos con las Escrituras tendrán una actitud diferente, nada proclives a este actualismo geológico. John Ray, por

---

15. A. Scilla, La vana speculazione disingannata dal senso. Lettera risponsiva circa i corpi marini, che petrificati si trovano in varij luoghi terrestri, op. cit., 108.

16. Ibidem, 13.

17. W. Harvey, Exercitationes de generatione animalium, Londres, 1651.

18. A. Scilla, La vana speculazione, op. cit., 11.

19. J.M. Artola, A. Galera, El tiempo biológico, op. cit., 226.

ejemplo, acepta la condición orgánica de los fósiles sin revocar el pasado creacionista que proclama el texto sagrado, y, como respuesta a la teoría de Scilla,
en sus *discourses* rechaza las inundaciones explicando la distribucción territorial de los fósiles mediante un acto divino que se remota a la época de la Creación. No son las aguas las que descienden de las montañas sino la tierra la que
emerge del fondo del mar, que originariamente cubría la Tierra. Montañas
donde se produjo la creación de los animales y el hombre : " Mas acorde, por
consiguiente, con las Escrituras y la razón es que al principio la Tierra estuvo
cubierta por el agua, que el suelo se levanto, por orden divina, mediante los
fuegos subterráneos, y aquí fueron gradualmente creados los animales y el
hombre, y después, poco a poco, las aguas fueron conducidas al fondo "[20]. El
naturalista inglés expone un modelo estático apropiado a su ideología fijista, la
fosilización es un acontecimiento lejano y primigenio, que no forma parte de
la historia reciente de una Tierra ajena a todo cambio.

El ideario de Scilla alcanzó el horizonte del setecientos con explendor, fue,
por ejemplo, elemento de controversia en la tribuna de la Royal Society londi-
nense, donde en 1696 William Wotton presentó un resumen de la obra[21]. La
atención prestada por John Ray en la tercera edición de sus *discourses*, deba-
tiendo sus argumentos[22] ; el testimonio de Leibniz en su *Protogaea*, reconociendo el acierto paleontológico del *savant peindre*[23] ; y el análisis realizado
por Benoit de Maillet en su *Telliamed*, exponiendo ampliamente la obra[24] ; le
hicieron partícipe de la polémica ilustrada sobre los fósiles[25]. Pero la Historia
no ha sido siempre tan favorable. Lyell, por ejemplo, en las pocas líneas que
ocupa Scilla en la semblanza histórica de los *Principles*, elogia su defensa de
la condición orgánica de los fósiles y recrimina una, supuesta, ideología dilu-
vista que consideraría a los fósiles " el efecto y la prueba del diluvio de
Moises "[26]. El equívoco denota ignorancia, indica el desconocimiento de una
obra y un ideario cuya principal hipótesis es que los fósiles son " verdaderos
animales, y no bromas de la Naturaleza generadas simplemente de sustancia

---

20. *Cf.* Ch.E. Raven, *John Ray, naturalist : his Life and works*, Cambridge, 1942, 438 (reimp. 1986).

21. *The Philosophical Transaction* publicó en el n° 219, enero y febrero de 1695, 181-199, un amplio resumen de la obra. En 1697 W. Wotton publicá en Londres *A vindication of an Abstract of an Italian Book concerning Marine Bodies*. *Cf.* también P. Rossi, *I segni del tempo*, Milán, 1979, 44.

22. Ch.E. Raven, *John Ray, naturalist, op. cit.*, 138-145. La correspondencia de Ray revela que ya en 1696 el naturalista inglés tenía conocimiento detallado del ideario paleontológico de Scilla ; *cf.* Ray, *Further correspondance*, Ray Society, 1928, 266.

23. Leibniz, *Protogée ou de la formation et des révolutions du globe*, Paris, 1859, 73 y 79.

24. B. de Maillet, *Telliamed ou entretiens d'un philosophe indien avec un missionnaire fran-çais*, Paris, 1748, 176-182 (reed. 1984).

25. En esta dirección, no debemos olvidar que la obra Scilla se reeditó en Roma en 1752 con el título *De corporibus marinis lapidescentibus*, edición que incluye el texto de Fabio Colonna *De Glossopetris*.

26. Ch. Lyell, *Principles of Geology*, I, *op. cit.*, 37.

pétrea "[27]. La procedencia es un problema secundario, que, en cualquier caso, merece una respuesta geofísica actualista : desecaciones, inundaciones y otros fenómenos naturales son el probable origen de la confusión[28].

Al margen del error nominal, Lyell acierta al cuestionar la licitud del consenso que dirime las diferencias entre ciencia y sociedad para que la nueva interpretación de la naturaleza, surgida al amparo del registro fósil, se asimile sin trasgredir radicalmente la norma social, impidiéndo que los científicos lleven a sus últimas consecuencias los datos observados. Se trata, simplemente, del tradicional debate entre ciencia y religión que, al menos a corto plazo, supone el triunfo de la fe sobre la razón, equivale a renunciar a los hechos : " porque de poca cosa servía que la naturaleza de los documentos fuese, por fin, bien comprendida, si los hombres eran impedidos de sacar las consecuencias exactas "[29]. La conclusión, en el caso que nos ocupa, era denunciar el tradicional modelo creacionista.

En la génesis de una paleontología la pregunta ¿ qué es la vida ? tiene una respuesta integradora, construida desde la relación de animales y plantas con el reino mineral. Si en la *Historia Natural* de Plinio, por ejemplo, esta broma de la naturaleza no es más que tierra petrificada, *lapidem vertuntur*, y el problema no sobrepasa el ámbito del reino mineral, es un componente más, en el Renacimiento napolitano se modula el concepto de fósil como una realidad orgánica fruto de una historia geológica de la Tierra reciente, y no de un remoto pasado creacionista. El fijismo y la perfección de la naturaleza rigen los destinos de un sistema natural que comienza a vocalizar la palabra cambio. Así, finalizando el siglo, Robert Hooke define una nueva frontera al hablar de " especies de criaturas qui no tienen representantes actuales "[30]. Hasta entonces los restos paleontológicos tenían su correspondencia con el presente ; para las formas desconocidas se podía apelar a la ignorancia, ubicarlas en algún remoto lugar del Globo como propuso John Ray.

27. A. Scilla, *La vana speculazione, op. cit.*, 32.

28. *Ibidem*, 129.

29. Ch. Lyell, *Principles of Geology, op. cit.*, p. 38.

30. R. Hooke, *The posthumous works, containing his Cutlerian lectures and other discourses*, Londres, 1705 (reimp. 1969).

CONTINU / DISCONTINU EN PALÉONTOLOGIE ÉVOLUTIVE

Goulven LAURENT

Le problème de la continuité de la vie sur la terre est un des problèmes fondamentaux — le problème fondamental — qui s'est posé à la théorie évolutionniste. C'est sans doute un problème très ancien, mais il est devenu un objet d'études et de débats, et même de polémiques, dès la première formulation scientifique de la transformation des êtres vivants les uns dans les autres au cours des temps, c'est-à-dire dès les premières années du XIXᵉ siècle.

Comme on le sait, c'est Lamarck (1744-1829) qui a lancé dans le monde des naturalistes, en 1800, la théorie transformiste[1]. L'affrontement Continu / Discontinu allait immédiatement devenir public dans le conflit entre lui et Cuvier (1769-1832).

On connaît les origines et les circonstances de ce débat. C'est Cuvier qui l'avait lancé, en affirmant que le passé avait été coupé à plusieurs reprises par des catastrophes " générales ", ou par des inondations " universelles "[2]. C'est sur la paléontologie qu'il se base pour soutenir le catastrophisme. Après avoir étudié les éléphants fossiles, et avoir rappelé le cas des rhinocéros, des ours et autres espèces fossiles, dont " aucun n'a d'analogue vivant ", Cuvier conclut en effet : " tous ces faits, analogues entre eux, et auxquels on n'en peut opposer aucun de constaté, me paraissent prouver l'existence d'un monde antérieur au nôtre, détruit par une catastrophe quelconque "[3]. En face de lui se dresse alors Lamarck, qui se base lui aussi sur la paléontologie pour réfuter la théorie de son collègue du Muséum. Lamarck est en effet, comme l'on sait, le fondateur de l'étude des invertébrés. C'est lui qui en a inventé le nom, en 1797[4], et

---

1. " Discours d'ouverture du cours de l'an VIII " (1800), *Bulletin scientifique de la France et de la Belgique*, 40 (1907), 459-482.

2. *Discours sur les révolutions de la surface du globe*, 1825, 62, 285, 292, 329, 330 (" dernière inondation universelle "), 332, 335.

3. " Mémoire sur les espèces d'Eléphans, tant vivantes que fossiles ", *Magasin encyclopédique*, 2ᵉ année, 3 (1796), 444.

4. *Mémoires de Physique et d'Histoire naturelle*, 1797, 314.

il a été le premier à y mettre de l'ordre, un ordre reconnu comme scientifique par ses contemporains. Il a été aussi le fondateur de leur paléontologie, en étudiant et en classant plus d'un millier de leurs fossiles. C'est en se basant aussi sur eux qu'il entreprend de réfuter la théorie discontinuiste, autrement dit catastrophiste, de Cuvier.

Dès le point de départ, il y a donc eu affrontement entre deux écoles de pensée : celle du fixiste Cuvier et celle du transformiste Lamarck, et mise en opposition de deux catégories de fossiles : les 170 espèces fossiles de vertébrés de Cuvier, et le millier d'espèces d'invertébrés de Lamarck. Cuvier avait utilisé ses fossiles vertébrés pour démontrer la discontinuité de la vie sur la terre, Lamarck utilise ses fossiles d'invertébrés pour en affirmer la continuité.

Le concept opératoire qu'il utilise est celui des fossiles " analogues ". Pour Cuvier, il n'y a pas de fossiles analogues : comme nous l'avons vu affirmer, toutes les espèces du passé sont différentes des espèces actuelles. Il ne peut en effet y avoir aucun rapport entre elles, car les espèces anciennes ont été détruites.

Pour Lamarck, au contraire, il y a une cinquantaine d'espèces anciennes qui sont " analogues " aux espèces d'aujourd'hui. Ces espèces fossiles sont donc en fait les mêmes espèces que les espèces actuelles, mais changées " par le temps et les circonstances "[5].

Lamarck formalise sa démonstration contre Cuvier, en montrant comment l'existence de telles espèces ruine le catastrophisme. Certains prétendent, écrit-il, et c'est évidemment Cuvier qui est visé, " que tous les fossiles appartiennent à des dépouilles d'animaux ou de végétaux dont les analogues vivans n'existent plus dans la nature ", et ils en ont conclu " que ce globe a subi un bouleversement universel, une catastrophe générale "[6]. Mais, argumente Lamarck, il suffit, pour ruiner cette thèse, de montrer que " quelques " espèces ont survécu. Un petit nombre de fossiles analogues suffit en effet, dès lors " qu'on ne sauroit le contester ", pour que l'on soit " forcé de supprimer l'universalité énoncée dans la proposition citée ci-dessus "[7]. C'est un des plus beaux exemples de démonstration que l'on possède en argumentation scientifique.

Mais Cuvier ne s'avoua pas vaincu pour autant. Il revint à la charge, et il formula une objection très sérieuse à la théorie transformiste de Lamarck et de ses disciples, selon laquelle les êtres actuels descendaient par voie ininterrompue de descendance des êtres anciens. " On peut, argumentait Cuvier, leur répondre, dans leur propre système, que si les espèces ont changé par degrés, on devroit trouver des traces de ces modifications graduelles ; qu'entre le palaeotherium et les espèces d'aujourd'hui l'on devroit découvrir quelques for-

---

5. " Mémoire sur les Fossiles des environs de Paris ", *Annales du Muséum d'Histoire naturelle de Paris*, 6 (1805), 220.

6. *Système des Animaux sans vertèbres*, 1801, 407.

7. *Idem*, 408.

mes intermédiaires ". Or, fait-il remarquer, " jusqu'à présent cela n'est point arrivé "[8]. Voilà le débat tel qu'il fut lancé, et l'argumentation employée, dès les premières années du XIXᵉ siècle. Nous allons le retrouver, dans les mêmes termes, dans les années qui vont suivre, … et même jusqu'à aujourd'hui.

Le débat Continu / Discontinu allait en effet se poursuivre activement après la mort de Cuvier et de Lamarck. La paléontologie des vertébrés continue à soutenir la discontinuité, non seulement en France, mais aussi en Angleterre, avec Buckland (1784-1856) en particulier, et en Suisse puis en Amérique, avec Louis Agassiz (1807-1873). Ce domaine ne se renouvellera que dans les années 1850, avec l'arrivée sur le devant de la scène de la jeune génération de naturalistes, en particulier Albert Gaudry (1827-1908), dont nous reparlerons. Ces jeunes ont été formés par Isidore Geoffroy Saint-Hilaire (1805-1861), qui leur a enseigné à considérer les êtres actuels, et ceux de chaque période, comme les descendants des êtres qui les ont précédés. Pour vérifier la justesse de la " théorie de la variabilité ", qui est " incontestablement la plus simple et la moins conjecturale "[9], il suffit, enseigne-t-il à ses étudiants, de faire la " comparaison des espèces actuelles avec celles de l'époque antérieure, ou, plus généralement, des espèces de deux époques consécutives, en vue d'établir leurs rapports de filiation "[10].

La discipline la plus vivante et la plus active dans le débat d'idées était évidemment la paléontologie des invertébrés. La classification de Lamarck était considérée par les invertébristes avec une sorte de " vénération ", comme nous l'apprend Gérard-Paul Deshayes[11]. Ils ont suivi sa problématique, et nous voyons ainsi ses idées anticatastrophistes faire rapidement leur chemin au cours de la première moitié du XIXᵉ siècle. Grâce à ces nombreux travaux réalisés dans ce domaine — les ouvrages de Lamarck à la main, il ne faut pas l'oublier — la notion de continuité se propage rapidement, au point que, avant même la mort de Cuvier, la majorité des géologues-paléontologistes se déclarent partisans de la continuité, et en opposition avec le catastrophisme discontinuiste de Cuvier.

André de Férussac (1786-1836) abondait dans le sens de Lamarck. S'il s'est spécialisé dans l'étude des invertébrés, et spécialement dans celle des mollusques, c'est qu'il est convaincu que c'est par eux que passe l'établissement de la véritable histoire de la terre et de la vie.

L'oeuvre de Defrance (1758-1850) s'articule aussi essentiellement sur les espèces " analogues ". C'est véritablement autour et à partir de ce concept qu'il a rassemblé son immense collection de fossiles.

---

8. " Discours préliminaire ", *Recherches sur les Ossemens fossiles de Quadrupèdes*, t. 1, 1812, 73.

9. *Histoire naturelle générale des Règnes organiques*, t. 2, 1859, 435.

10. *Idem*, 437.

11. Lettre à l'éditeur de *Zeitschrift für Malakozoologie*, année 1845, 45.

De Basterot (mort en 1887) s'est aussi beaucoup intéressé aux mollusques. Il les étudie sous le rapport de l'analogie, et observe les rapprochements à faire. Les espèces fossiles et actuelles ne sont, à ses yeux, que " des modifications d'une seule espèce réelle produites par les circonstances "[12]. Lamarck n'avait pas dit autre chose.

Marcel de Serres (1780-1862) se trouvait lui aussi en accord avec Lamarck, en désaccord par conséquent avec Cuvier. La marche de la nature a été continue," le fil de ses opérations n'est pas brisé "[13], affirme-t-il, rappelant la formule célèbre de Cuvier en la contredisant.

Ami Boué (1794-1881) refusait aussi le catastrophisme de Cuvier. " Malgré les talens zoologiques de M. Cuvier, écrit-il en 1831, la pluralité des géologues sont d'accord pour regarder le déluge mosaïque, pris à la lettre, comme un des événemens géologiques les moins prouvés "[14]. Ami Boué ne croyait pas davantage à l'idée " ancienne "[15] de la fixité des espèces qu'il ne croyait aux catastrophes. Dans cette question, il préférait lui aussi suivre " les Lamarck, les Geoffroy et autres grands naturalistes "[16].

L'idée de la continuité de la vie, infusée par ce développement qui semblait régulier dans sa progression, renforçait ainsi la crédibilité de la doctrine de Lamarck. " Aujourd'hui ", écrit Boblaye en 1834 — rappelons que Cuvier n'est mort que deux ans plus tôt — " que l'étude des animaux fossiles a beaucoup étendu les limites des variations attribuées aux espèces, et fait découvrir chaque jour dans la chaîne des êtres des transitions qu'on ne soupçonnait pas, l'hypothèse hardie de Lamarck, modifiée par Geoffroy Saint-Hilaire, acquiert une probabilité qu'elle n'avait pas à l'époque où Cuvier la combattait "[17].

Pour un paléontologiste comme Théophile Ebray (1823-1879), qui a beaucoup étudié les terrains de l'Ouest de la France, les séries de fossiles qu'il établit d'une manière de plus en plus rigoureuse, apportaient justement une évidence irréfutable de la transformation de leurs types primitifs.

Il n'est pas étonnant que grâce à ces travaux sur les fossiles, continuateurs de ceux de Lamarck, l'histoire du XIXe siècle ne soit jalonnée, dès le temps même de Lamarck, et continuellement après lui, de proclamations transformistes.

12. " Description géologique du Bassin tertiaire du Sud-Ouest de la France ", *Mémoires de la Société d'Histoire naturelle de Paris*, 2 (1825), 7-8, dans la note.

13. " Observations sur la pétrification des coquilles dans la Méditerranée ", *Revue scientifique et industrielle*, 2e série, 14 (1847), 395.

14. *Bulletin des Sciences naturelles et de Géologie*, t. 26 (1831), 2-3.

15. " Résumé des progrès des sciences géologiques ", *Bulletin de la Société géologique de France*, 5 (1834), 113.

16. *Idem*, 114.

17. " Animaux fossiles ", *Dictionnaire pittoresque d'Histoire naturelle et des phénomènes de la Nature*, 1 (1834), 192.

Aussi nous ne sommes pas étonnés de voir Frédéric Gérard (1806-1857) parler de " la théorie de l'évolution des formes organiques "[18] ; du " mouvement évolutif "[19] ; de la " loi d'évolution "[20] ; de " l'évolution des êtres organisés "[21] ; de " périodes évolutives "[22], et même soutenir proprement, en 1847 !, " la doctrine de l'évolution " !...[23].

Camille Dareste, enfin, en 1859, à la veille de la parution de l'*Origine des espèces* de Darwin, constate que l'on voit " les hommes les plus éminents entrer dans la voie ouverte par Lamarck, et faire de l'idée de la variabilité limitée des espèces le point de départ de leurs théories scientifiques "[24].

Cependant, d'Omalius d'Halloy (1783-1875), le fondateur de la géologie belge, qui fut, dès 1831, parmi les géologues et les paléontologistes, l'un des premiers et des plus ardents défenseurs du transformisme, souligne le fait que, dès le début de ce qui est connu par la paléontologie, tous les grands types organiques se trouvaient déjà représentés dans la nature[25]. De Saporta (1823-1895), paléobotaniste du midi de la France, a passé pour un héraut du transformisme, et l'un de ses plus chauds partisans. De Saporta est, cependant, profondément conscient de la distance qui sépare les hypothèses imaginées des faits constatés. Ainsi, en ce qui concerne les premières dicotylédones, il avoue que c'est seulement " théoriquement " qu'il peut se les " figurer ", et que l'" état primitif " caractéristique de " la grande simplification organique " qu'il leur attribue ne saurait être définie " que d'une façon conjecturale "[26]. Il ne rejette même pas l'éventualité de la fausseté de cette représentation, car " cette structure idéale primitive " n'a peut-être pas existé[27].

Oswald Heer (1809-1883), paléobotaniste suisse, est transformiste lui aussi[28]. Mais il est obligé de constater que les relations entre les espèces se font sur le mode brusque, ce qui veut dire que toutes les explications qui se fondent sur la transformation lente sont infirmées par les fossiles. La paléontologie montre " qu'il suffit d'un temps relativement peu considérable pour qu'une espèce puisse se modeler sous toutes ses formes possibles, et s'adapter aux circonstances extérieures pour rester ensuite immobile pendant des milliers

---

18. Article " Dégénérescence ", *Dict. Univ. Hist. nat.*, 4 (1844), 649.
19. Article " Géographie zoologique ", *Dict. Univ. Hist. nat.*, 6 (1845), 112 ; *cf.* encore 119.
20. Article " Espèce ", *loc. cit.*, 5 (1844), 434.
21. *Idem*, 432.
22. Article " Géographie zoologique ", *loc. cit.*, 6 (1845), 112.
23. " De la finalité : inconciliabilité de cette doctrine avec la philosophie naturelle ", *Rev. scient. industr.*, (2) 13 (1847), 372.
24. Biographie de Lamarck, *Nouvelle Biographie générale*, t. 29, de Hoefer, 1859, 62.
25. *Bulletin de la Société géologique de France*, 2, 16 (1858-9), 515.
26. " Etudes sur la végétation du Sud-Est de la France à l'époque tertiaire ", *Annales des Sciences naturelles. Botanique*, 5ᵉ série, 4 (1865), 16.
27. *Ibidem*.
28. *Cf.* l'étude de B. Hoppe " Die Evolution der Organismen im Denken des Paläontologen Oswald Heer (1809-1883) ", *Medizin historisches Journal*, Band 19, Heft 4 (1984), 348-362.

d'années "[29]. Il faut donc bien distinguer la période de stabilité des plantes de la période de leur " remaniement ".

Eduard Suess (1831-1914), né à Londres, et professeur à Vienne, en Autriche, est connu par son maître ouvrage *Das Antlitz der Erde*. Il y développe ses convictions transformistes, qu'il avait déjà affirmées très tôt, dans des articles de jeunesse publiés dès les années 1850. Mais lui non plus ne peut pas ne pas constater que les documents paléontologiques ne confirment pas le principe de base de la théorie évolutionniste, car la transformation des espèces ne se fait pas graduellement. " Nous ne voyons pas, assure-t-il, au sein d'une même famille ou d'un même genre, les espèces se modifier d'une manière continue et à des époques différentes : au contraire, ce sont des associations entières, toutes les populations animales et végétales, les grandes unités de l'économie de la nature… qui font leur apparition ou disparaissent simultanément "[30].

Karl von Zittel (1839-1904) soutient que " chaque ensemble faunique est apparu par la transformation graduelle de ses éléments à partir des éléments d'une faune précédente, et il a fourni de la même façon les éléments de l'ensemble faunique suivant "[31]. Mais il constate qu'il " existe avec évidence des périodes où le processus de changement et la destruction des formes organiques se sont réalisées d'une manière particulièrement rapide et énergique, et il existe entre ces périodes de changement de longues périodes de repos pendant lesquelles les espèces conservent leurs caractères particuliers à peu près inchangés "[32].

Albert Gaudry (1827-1908) a été considéré comme le père de la paléontologie évolutive, et Darwin lui accordait beaucoup de confiance. C'est " un paléontologiste distingué "[33], assurait-il, et il prédisait que Gaudry " sera bientôt un des principaux chefs en matière de Paléontologie zoologique en Europe "[34].

Mais Gaudry reconnaît que " ceux mêmes des reptiles primaires où l'on observe des caractères d'infériorité n'établissent pas de liens entre la classe des reptiles (allantoïdiens ou anallantoïdiens) et celle des poissons "[35].

Un tel consensus dans la pensée et une telle continuité dans l'expression chez les paléontologues post-darwiniens ne peuvent être sans signification ni surtout sans portée. Elles montrent que les fossiles ont toujours été des empêcheurs de tourner intellectuellement en rond. Le lecteur aura compris en tout

---

29. " Discours prononcé à l'ouverture de la 48ᵉ session de la Société helvétique des Sciences naturelles ", *Annales des Sciences naturelles. Botanique*, 5ᵉ série, 3 (1865), 185-186.

30. *La Face de la Terre*, t. 1, 1897, 15.

31. *Grundzüge der Paläontologie*, 1895, 948.

32. *Idem*, 15.

33. *The Origin of Species*, (1ʳᵉ édit., 1859), 6ᵉ édit., 1872, 301.

34. *La vie et la correspondance de Charles Darwin*, t. 2, traduct. de Varigny, 1888, 418.

35. *Les Enchaînements du Monde animal. Fossiles primaires*, 1883, 287-288.

cas que la reprise actuelle des mêmes thèmes et des mêmes constatations par Elredge et Gould est plutôt la réaffirmation d'une position séculaire que la découverte d'une nouveauté révolutionnaire. La conviction scientifique des paléontologistes est restée en effet étonnamment la même, à travers les vicissitudes de l'envahissement (ou de l'affaiblissement) d'idéologies à prétention totalitaire.

On ne peut que souhaiter que les paléontologistes continuent à travailler, et à nous apporter de vieux os et de l'air frais !

THE LATER DISCOVERIES OF BOUCHER DE PERTHES AT MOULIN QUIGNON AND THEIR BEARING ON THE MOULIN QUIGNON JAW CONTROVERSY

Michael A. CREMO

My book *Forbidden Archaeology*[1], co-authored with Richard L. Thompson, examines the history of archaeology and documents numerous discoveries suggesting that anatomically modern humans existed in times earlier than now thought likely. According to most current accounts, anatomically modern humans emerged within the past one or two hundred thousand years from more primitive ancestors. Much of the evidence for greater human antiquity, extending far back into the Tertiary, was discovered by scientists in the 19th and early 20th centuries. Current workers are often unaware of this remarkable body of evidence. In their review article about *Forbidden Archaeology*, historians of science Wodak and Oldroyd[2] suggest that "perhaps historians bear some responsibility" for this lack of attention.

"Certainly, some pre-FA histories of palaeoanthropology, such as Peter Bowler's, say little about the kind of evidence adduced by C&T, and the same may be said of some texts published since 1993, such as Ian Tattersall's recent book. So perhaps the rejection of Tertiary [and early Pleistocene] *Homo sapiens*, like other scientific determinations, is a social construction in which historians of science have participated. C&T claim that there has been a "knowledge filtration operating within the scientific community", in which historians have presumably played their part".

I am also guilty of this knowledge filtering. In this paper, I give an example of my own failure to free myself from unwarranted prejudice.

---

1. M.A. Cremo, R.L. Thompson, *Forbidden Archaeology : The Hidden History of the Human Race*, San Diego, 1993.
2. J. Wodak, D. Oldroyd, "Vedic Creationism : A Further Twist to the Evolution Debate", *Social Studies of Science*, 26 (1996), 192-213, esp. 197.

The collective failure of scientists and historians to properly comprehend and record the history of investigations into human antiquity has substantial consequences on the present development of human antiquity studies. Current workers should have ready access to the complete data set, not just the portion marshalled in support of the current picture of the past and the history of this picture's elaboration. The value of historians' work in maintaining the complete archive of archaeological data in accessible form can thus be significant for ongoing human antiquity studies. This approach does not, as some have suggested, entail uncritical acceptance of all past reporting. But it does entail suspension of naive faith in the progressive improvement in scientific reporting.

In February of 1997, I lectured on *Forbidden Archaeology* to students at the faculty of archaeology and earth sciences at the University of Louvain, Belgium. Afterwards, one of the students, commenting on some of the 19th century reports I presented, asked how we could accept them, given that these reports had already been rejected long ago and that scientific understanding and methods had greatly improved since the 19th century. I answered, " If we suppose that in earlier times scientists accepted bad evidence because of their imperfect understanding and methods then we might also suppose they rejected good evidence because of their imperfect understanding and methods. There is no alternative to actually looking critically at specific cases ". This is not necessarily the task of the working archaeologist. But the historian of archaeology may here play a useful role.

The specific case I wish to consider is that of the discoveries of Jacques Boucher de Perthes at Moulin Quignon. This site is located at Abbeville, in the valley of the Somme in north-eastern France. In *Forbidden Archaeology*, I confined myself to the aspects of the case that are already well known to historians[3]. To summarize, in the 1840s Boucher de Perthes discovered stone tools in the Middle Pleistocene high level gravel's of the Somme, at Moulin Quignon and other sites. At first, the scientific community, particularly in France, was not inclined to accept his discoveries as genuine. Some believed that the tools were manufactured by forgers. Others believed them to be purely natural forms that happened to resemble stone tools. Later, leading British archaeologists visited the sites of Boucher de Perthes's discoveries and pronounced them genuine. Boucher de Perthes thus became a hero of science. His discoveries pushed the antiquity of man deep into the Pleistocene, coeval with extinct mammals. But the exact nature of the maker of these tools remained unknown. Then in 1863, Boucher de Perthes discovered at Moulin Quignon additional stone tools and an anatomically modern human jaw. The jaw inspired much controversy, and was the subject of a joint English-French commission. To do justice to the

3. M.A. Cremo, R.L. Thompson, *Forbidden Archaeology : The Hidden History of the Human Race, op. cit.*, 402-404.

entire proceedings[4] would take a book, so I shall in this paper touch on only a few points of contention.

The English members of the commission thought the recently discovered stone tools were forgeries that had been artificially introduced into the Moulin Quignon strata. They thought the same of the jaw. To settle the matter, the commission paid a surprise visit to the site. Five flint implements were found in the presence of the scientists. The commission approved by majority vote a resolution in favor of the authenticity of the recently discovered stone tools. Sir John Prestwich remained in the end sceptical but nevertheless noted[5] that " the precautions we took seemed to render imposition on the part of the workmen impossible ".

That authentic flint implements should be found at Moulin Quignon is not surprising, because flint implements of unquestioned authenticity had previously been found there and at many other sites in the same region. There was no dispute about this at the time, nor is there any dispute about this among scientists today. The strange insistence on forgery and planting of certain flint implements at Moulin Quignon seems directly tied to the discovery of the Moulin Quignon jaw, which was modern in form. If the jaw had not been found, I doubt there would have been any objections at all to the stone tools that were found in the gravel pit around the same time.

In addition to confirming the authenticity of the stone tools from Moulin Quignon, the commission also concluded that there was no evidence that the jaw had been fraudulently introduced into the Moulin Quignon gravel deposits[6]. The presence of grey sand in the inner cavities of the jaw, which had been found in a blackish clay deposit, had caused the English members of the commission to suspect that the jaw had been taken from somewhere else. But when the commission visited the site, some members noted the presence of a layer of fine grey sand just above the layer of black deposits in which the jaw had been found[7]. This offered an explanation for the presence of the grey sand in the Moulin Quignon jaw and favored its authenticity.

Trinkaus and Shipman[8] insinuate, incorrectly, that the commission's favor-

4. H. Falconer, G. Busk, W.B. Carpenter, " An account of the proceedings of the late conference held in France to inquire into the circumstances attending the asserted discovery of a human jaw in the gravel at Moulin-Quignon, near Abbeville ; including the *procès verbaux* of the conference, with notes thereon ", *The Natural History Review*, 3 (new series) (1863), 423- 462 ; A. Delesse, " La mâchoire humaine de Moulin de Quignon ", *Mémoires de la Société d'Anthropologie de Paris*, 2 (1863), 37-68.

5. J. Prestwich, " On the section at Moulin Quignon Abbeville, and on the peculiar character of some of the flint implements recently discovered there ", *Quarterly Journal of the Geological Society of London*, vol. 19, first part (1863), 497-505, esp. 505.

6. H. Falconer, G. Busk, W.B. Carpenter, " An account of the proceedings of the late conference held in France... ", *op. cit.*, 452.

7. *Idem*, 448-449.

8. E. Trinkaus, P. Shipman, *The Neanderthals*, New York, 1992, 96.

able resolution simply absolved Boucher de Perthes of any fraudulent introduction of the jaw (hinting that others may have planted it). But that is clearly not what the commission intended to say, as anyone can see from reading the report in its entirety.

Here are the exact words of Trinkaus and Shipman : " In any case, the commission found itself deadlocked. There was only one point of agreement : " The jaw in question was not fraudulently introduced into the gravel pit of Moulin Quignon [by Boucher de Perthes] ; it had existed previously in the spot where M. Boucher de Perthes found it on the 28th March 1863 ". This lukewarm assertion of his innocence, rather than his correctness, was hardly the type of scientific acclaim and vindication that Boucher de Perthes yearned for " [the interpolation is by Trinkaus and Shipman].

But the commission[9] also voted in favor of the following resolution : " All leads one to think that the deposition of this jaw was contemporary with that of the pebbles and other materials constituting the mass of clay and gravel designated as the black bed, which rests immediately above the chalk ". This was exactly the conclusion desired by Boucher de Perthes. Only two members, Busk and Falconer, abstained. The committee as a whole was far from deadlocked.

Their scientific objections having been effectively countered, the English objectors, including John Evans, who was not able to join the commission in France, were left with finding further proof of fraudulent behavior among the workmen at Moulin Quignon as their best weapon against the jaw. Taking advantage of a suggestion by Boucher de Perthes himself, Evans sent his trusted assistant Henry Keeping, a working man with experience in archaeological excavation, to France. There he supposedly obtained definite proof that the French workmen were introducing tools into the deposits at Moulin Quignon.

But careful study of Keeping's reports[10] reveals little to support these allegations and suspicions. Seven implements, all supposedly fraudulent, turned up during Keeping's brief stay at Moulin Quignon. Five were found by Keeping himself and two were given to him by the two French workers who were assigned by Boucher de Perthes to assist him. Keeping's main accusation was that the implements appeared to have " fingerprints " on them. The same accusation had been levelled by the English members of the commission against the tools earlier found at Moulin Quignon. In his detailed discussion of Keeping, which is well worth reading, Boucher de Perthes[11] remarked that he and others

---

9. H. Falconer, G. Busk, W.B. Carpenter, " An account of the proceedings of the late conference held in France... ", *op. cit.*, 452.

10. J. Evans, " The human remains at Abbeville ", *The Athenaeum* (July 4, 1863), 19-20.

11. J. Boucher de Perthes, " Fossile de Moulin-Quignon : Vérification Supplémentaire ", dans J. Boucher de Perthes (éd.), *Antiquités Celtiques et Antédiluviennes. Mémoire sur l'Industrie Primitive et les Arts à leur Origine*, vol. 3, Paris, 1864, 207-208.

had never been able to discern these fingerprints. Boucher de Perthes[12] also observed that Keeping was daily choosing his own spots to work and that it would have been quite difficult for the workers, if they were indeed planting flint implements, to anticipate where he would dig. I tend to agree with Boucher de Perthes[13] that Keeping, loyal to his master Evans, was well aware that he had been sent to France to find evidence of fraud and that he dared not return to England without it. Evans's report[14], based on Keeping's account, was published in an English periodical and swayed many scientists to the opinion that Boucher de Perthes was, despite the favorable conclusions of the scientific commission, the victim of an archaeological fraud.

Not everyone was negatively influenced by Keeping's report. In *Forbidden Archaeology*, I cited Sir Arthur Keith[15], who stated, " French anthropologists continued to believe in the authenticity of the jaw until between 1880 and 1890, when they ceased to include it in the list of discoveries of ancient man ".

I also was inclined to accept the jaw's authenticity, but given the intensity of the attacks by the English, in *Forbidden Archaeology* I simply noted, " From the information we now have at our disposal, it is difficult to form a definite opinion about the authenticity of the Moulin Quignon jaw ". I stated this as a mild antidote to the nearly universal current opinion that the Moulin Quignon jaw and accompanying tools were definitely fraudulent. But because Evans and his English accomplices had so thoroughly problematized the evidence, I could not bring myself to suggest more directly that the Moulin Quignon jaw was perhaps genuine.

Boucher de Perthes, however, entertained no doubts as to the authenticity of the jaw, which he had seen in place in the black layer toward the bottom of the Moulin Quignon pit. He believed it had been rejected because of political and religious prejudice in England. Stung by accusations of deception, he proceeded to carry out a new set of excavations, which resulted in the recovery of more human skeletal remains. These later discoveries are hardly mentioned in standard histories, which dwell upon the controversy surrounding the famous Moulin Quignon jaw.

For example, the later discoveries of Boucher de Perthes rate only a line or two in Grayson[16] : " Evans's demonstration of fraud and the strongly negative reaction of the British scientists ensured that the Moulin Quignon mandible would never be accepted as an undoubted human fossil. The same applied to additional human bones reported from Moulin Quignon in 1864 ".

12. J. Boucher de Perthes, " Fossile de Moulin-Quignon : Vérification Supplémentaire ", *op. cit.,* 197, 204.

13. *Idem*, 194-195.

14. J. Evans, " The human remains at Abbeville ", *op. cit.*

15. A. Keith, *The Antiquity of Man*, Philadelphia, 1928, 271.

16. D.K. Grayson, *The Establishment of Human Antiquity*, New York, 1983, 217.

Trinkaus and Shipman[17] are similarly dismiss : " Desperately, Boucher de Perthes continued to excavate at Moulin Quignon. He took to calling in impromptu commissions (the mayor, stray geology professors, local doctors, lawyers, librarians, priests, and the like) to witness the event when he found something, or thought he was about to find something, significant... Soon, the English and French scientists stopped coming to look at his material or paying any real attention to his claims ".

Although aware of these later discoveries, I did not discuss them in *Forbidden Archaeology*. I thus implicated myself in the process of inadvertent suppression of anomalous evidence posited in *Forbidden Archaeology*[18] : " This evidence now tends to be extremely obscure, and it also tends to be surrounded by a neutralizing nimbus of negative reports, themselves obscure and dating from the time when the evidence was being actively rejected. Since these reports are generally quite derogatory, they may discourage those who read them from examining the rejected evidence further ".

The cloud of negative reporting surrounding the Moulin Quignon jaw influenced not only my judgement of this controversial find but also discouraged me from looking into the later discoveries of Boucher de Perthes. So let us now look into these discoveries and see if they are really deserving of being totally ignored or summarily dismissed.

Boucher de Perthes[19], stung by the accusations he had been deceived, carried out his new investigations so as to effectively rule out the possibility of deception by workmen. First of all, they were carried out during a period when the quarry at Moulin Quignon was shut down and the usual workmen were not there[20]. Also, Boucher de Perthes made his investigations unannounced and started digging at random places. He would usually hire one or two workers, whom he closely supervised. Furthermore, he himself would enter into the excavation and break up the larger chunks of sediment with his own hands. In a few cases, he let selected workers, who were paid only for their labor, work under the supervision of a trusted assistant. In almost all cases, witnesses with scientific or medical training were present. In some cases, these witnesses organized their own careful excavations to independently confirm the discoveries of Boucher de Perthes.

Here follow excerpts from accounts by Boucher de Perthes and others of these later discoveries. They are taken from the proceedings of the local

17. E. Trinkaus, P. Shipman, *The Neanderthals, op. cit.*, 96.

18. M.A. Cremo, R.L. Thompson, *Forbidden Archaeology : The Hidden History of the Human Race, op. cit.*, 28.

19. J. Boucher de Perthes, " Nouvelles Découvertes d'Os Humains dans le Diluvium, en 1863 et 1864, par M. Boucher de Perthes. Rapport à la Société Impériale d'Emulation ", dans J. Boucher de Perthes (éd.), *Antiquités Celtiques et Antédiluviennes. Mémoire sur l'Industrie Primitive et les Arts à leur Origine*, vol. 3, Paris, 1864, 215-250.

20. *Idem*, 219.

Société d'Emulation. Most French towns had such societies, composed of educated gentlemen, government officials, and businessmen.

On April 19, 1864 Boucher de Perthes took a worker to the gravel pit, and on the exposed face of the excavation pointed out some places for a worker to dig. Boucher de Perthes[21] " designated every spot where he should strike with his pick ". In this manner, he discovered a hand axe, two other smaller worked flints, and several flint flakes. Then the worker's pick " struck an agglomeration of sand and gravel, which broke apart, as did the bone it contained ". Boucher de Perthes[22] stated " I took from the bank the part that remained, and recognized the end of a human femur ". This find occurred at a depth of 2.3 meters, in the hard, compacted bed of yellowish brown sand and gravel lying directly above the chalk. In this, as in all cases, Boucher de Perthes had checked very carefully to see that the deposit was undisturbed and that there were no cracks or fissures through which a bone could have slipped down from higher levels[23]. Digging further at the same spot, he encountered small fragments of bone, including an iliac bone, 40 centimeters from the femur and in the same plane[24].

On April 22, Boucher de Perthes found a piece of human skull 4 centimeters long in the yellow brown bed. This yellow-brown bed contains in its lower levels some seams of yellow-grey sand. In one of these seams, Boucher de Perthes found more skull fragments and a human tooth[25].

On April 24, Boucher de Perthes was joined by Dr. J. Dubois, a physician at the Abbeville municipal hospital and a member of the Anatomical Society of Paris. They directed the digging of a worker in the yellow-brown bed. They uncovered some fragments too small to identify. But according to Dubois they displayed signs of incontestable antiquity. Boucher de Perthes and Dubois continued digging for some time, without finding anything more. " Finally, " stated Boucher de Perthes[26], " we saw in place, and Mr. Dubois detached himself from the bank, a bone that could be identified. It was 8 centimeters long. Having removed a portion of its matrix, Mr. Dubois recognized it as part of a human sacrum. Taking a measurement, we found it was lying 2.6 meters from the surface ". About 40 centimeters away, they found more bones, including a phalanx. They then moved to a spot close to where the jaw was discovered in 1863. They found parts of a cranium and a human tooth, the latter firmly

21. J. Boucher de Perthes, " Nouvelles Découvertes d'Os Humains dans le Diluvium,... ", *op. cit.*, 219.

22. *Ibidem.*

23. *Ibidem.*

24. *Ibidem.*

25. *Idem*, 220.

26. *Idem*, 221.

embedded in a pebbly mass of clayey sand[27]. The tooth was found at a depth of 3.15 meters from the surface[28].

On April 28, Boucher de Perthes began a deliberate search for the other half of the sacrum he had found on April 24. He was successful, locating the missing half of the sacrum bone about 1 meter from where the first half had been found. He also found a human tooth fragment in a seam of grey sand. Studying the edge of the break, Boucher de Perthes noted it was quite worn, indicating a degree of antiquity[29].

On May 1, accompanied for most of the day by Dr. Dubois, Boucher de Perthes found three fragments of human skulls, a partial human tooth, and a complete human tooth[30]. On May 9, Boucher de Perthes[31] found two human skull fragments, one fairly large (9 centimeters by 8 centimeters).

On May 12, Boucher de Perthes carried out explorations in the company of Mr. Hersent-Duval, the owner of the Moulin Quignon gravel pit. They first recovered from the yellow bed, at a depth of about 2 meters, a large piece of a human cranium, 8 centimeters long and 7 centimeters wide. "An instant later", stated Boucher de Perthes[32], "the pick having detached another piece of the bank, Mr. Hersent-Duval opened it and found a second fragment of human cranium, but much smaller. It was stuck so tightly in the mass of clay and stones that it took much trouble to separate it".

On May 15, Boucher de Perthes extracted from one of the seams of grey sand in the yellow-brown bed, at a depth of 3.2 meters, a human tooth firmly embedded in a chunk of sand and flint. The tooth was white. Boucher de Perthes[33] noted : "It is a very valuable specimen, that replies very well to the ... objection that the whiteness of a tooth is incompatible with its being a fossil". He then found in the bed of yellow-brown sand "a human metatarsal, still attached in its matrix, with a base of flint"[34]. In the same bed he also found many shells, which also retained their white color. Boucher de Perthes[35] observed : "Here the color of the bank, even the deepest, does not communicate itself to the rolled flints, nor to the shells, nor to the teeth, which all preserve their native whiteness". This answered an earlier objection to the antiquity of the original Moulin Quignon jaw and a detached tooth found along with it.

27. J. Boucher de Perthes, " Nouvelles Découvertes d'Os Humains dans le Diluvium,... ", *op. cit.*, 222.
28. *Idem*, 223.
29. *Ibidem*.
30. *Ibidem*.
31. *Idem*, 223-224.
32. *Idem*, 224.
33. *Idem*, 225.
34. *Ibidem*.
35. *Idem*, 226.

On June 6, Boucher de Perthes[36] found in the yellow-brown bed, at a depth of 4 meters, the lower half of a human humerus, along with several less recognizable bone fragments. On June 7, he recovered part of a human iliac bone at the same place[37]. On June 8 and 9, he found many bone fragments mixed with flint tools, including many hand axes. Later on June 10, he returned with three workers to conduct bigger excavations. He found two fragments of tibia (one 14 centimeters long) and part of a humerus[38]. These bones had signs of wear and rolling. They came from a depth of 4 meters in the yellow-brown bed. Please note that I am just recording the discoveries of human bones. On many days, Boucher de Perthes also found fragments of bones and horns of large mammals. Boucher de Perthes[39] noted that the human bones were covered with a matrix of the same substance as the bed in which they were found. When the bones were split, it was found that traces of the matrix were also present in their internal cavities[40]. Boucher de Perthes[41] noted that these are not the kinds of specimens that could be attributed to " cunning workers ". On this particular day, Boucher de Perthes left the quarry for some time during the middle of the day, leaving the workers under the supervision of an overseer. Boucher de Perthes[42] then reported :

" In the afternoon, I returned to the bank. My orders had been punctually executed. My representative had collected some fragments of bone and worked flints. But a much more excellent discovery had been made — this was a lower human jaw, complete except for the extremity of the right branch and the teeth.

My first concern was to verify its depth. I measured it at 4.4 meters, or 30 centimeters deeper than the spot where I had that morning discovered several human remains. The excavation, reaching the chalk at 5.1 meters, faced the road leading to the quarry. It was 20 meters from the point, near the mill, where I found the half-jaw on March 28, 1863.

The jaw's matrix was still moist and did not differ at all from that of all the other bones from that same bed. The matrix was very sticky, mixed with gravel and sometimes with pieces of bone, shells, and even teeth.

The teeth were missing from the jaw. They were worn or broken a little above their sockets, such that the matrix that covered them impaired their recognition. The deterioration was not recent, but dated to the origin of the bank.

36. J. Boucher de Perthes, " Nouvelles Découvertes d'Os Humains dans le Diluvium,... ", *op. cit.*, 230.
37. *Idem*, 231.
38. *Ibidem.*
39. *Idem*, 232.
40. *Ibidem.*
41. *Ibidem.*
42. *Idem*, 233-235.

Although I did not see that jaw *in situ*, after having minutely verified the circumstances of its discovery, I do not have the least doubt as to its authenticity. Its appearance alone suffices to support that conviction. Its matrix, as I have said, is absolutely identical to that of all the other bones and flints from the same bed. Because of its form and hardness, it would be impossible to imitate.

The worker in the trench, after having detached some of the bank, took it out with his shovel. But he did not see the jaw, nor could he have seen it, enveloped as it was in a mass of sand and flint that was not broken until the moment that the shovel threw it into the screen. It is then ... that it was seen by the overseer.

He recognized it as a bone, but not seeing the teeth, he did not suspect it was a jaw. Mr. Hersent-Duval who happened to come by at that moment, was undeceived. He signalled the workers and told them to leave it as it was, in its matrix, until my arrival, which came shortly thereafter.

After a short examination, I confirmed what Mr. Hersent had said. It was not until then that the workers believed. Until that moment, the absence of teeth and the unusual form of the piece, half-covered with clay, had caused even my overseer himself to doubt.

I therefore repeat : here one cannot suspect anyone. Strangers to the quarry and the town, these diggers had no interest in deception. I paid them for their work, and not for what they found... Dr. Dubois, to whom I was eager to show it, found it from the start to have a certain resemblance to the one found on March 28, 1863 ".

On June 17, Hersent-Duval had some workers dig a trench. They encountered some bones. Hersent-Duval ordered them to stop work, leaving the bones in place. He then sent a message for Boucher de Perthes to come. Boucher de Perthes arrived, accompanied by several learned gentlemen of Abbeville, including Mr. Martin, who was a professor of geology, and also a parish priest. Boucher de Perthes[43] stated :

" Many fragments, covered in their matrix, lay at the bottom of the excavation, at a depth of 4 meters. At 3 meters, one could see two points, resembling two ends of ribs.

Mr. Martin, who had descended with us into the trench, touched these points, and not being able to separate them, thought that they might belong to the same bone. I touched them in turn, as did the priest Dergny, and we agreed with his opinion.

Before extracting it, these gentlemen wanted to assure themselves about the state of the terrain. It was perfectly intact, without any kind of slippage, fissures, or channels, and it was certainly undisturbed. Having acquired this cer-

---

43. J. Boucher de Perthes, " Nouvelles Découvertes d'Os Humains dans le Diluvium,... ", *op. cit.*, 235-237.

tainty, the extraction took place by means of our own hands, without the intermediary of a worker.

Mr. Martin, having removed part of the envelope of the extracted bone, recognized it as a human cranium. And the two points at first taken as two ends of ribs, were the extremities of the brow ridges. This cranium, of which the frontal and the two parietals were almost complete, astonished us with a singular depression in its upper part.

This operation accomplished, we occupied ourselves with the bones fallen to the bottom of the quarry. They were three in number, covered by a mass of clay so thick that one could not tell the kind of creature to which they belonged. Much later, they were identified by Dr. Dubois as a human iliac bone, a right rib, and two pieces of an upper jaw, perhaps from the same head as the partial cranium, because they came from the same bed.

Having continued our excavation, we found yet another human bone, and we probably would have encountered others, if we had been able, without the danger of a landslide, to carry out the excavation still further.

All of this was recorded by Abbé Dergny, in a report signed by him and professor Martin... one of the most knowledgeable and respected men of our town ".

On July 9th, a commission composed of the following individuals made an excavation at Moulin Quignon : Louis Trancart, mayor of Laviers ; Pierre Sauvage, assistant to the mayor of Abbeville, and member of the *Société d'Emulation* of that town ; F. Marcotte, conservator of the museum of Abbeville, and member of the *Société d'Emulation* and the Academy of Amiens, A. de Caïeu, attorney, and member of the *Société d'Emulation* and the Society of Antiquaries of Picardy ; and Jules Dubois, M.D., doctor at the municipal hospital of Abbeville, member of many scientific societies[44].

At the quarry they carried out excavations at two sites. Marcotte, who had proclaimed his scepticism about the discoveries, was chosen to direct the digging of the workers. " He had the base of the excavation cleared away until it was possible to see the chalk, upon which directly rested the bed of yellow-brown sand ", said Dubois[45] in his report on the excavation of the first site in the quarry. " After we assured ourselves that the wall of the cut was clearly visible to us and that it was free of any disturbance, the work commenced under our direct inspection ". After 15 minutes of digging, Marcotte recovered a bone that Dubois[46] characterized as probably a piece of a human radius 8 centime-

44. J. Dubois, " Untitled report of excavation at Moulin Quignon, on July 9, 1864. Société Impériale d'Emulation. Extrait du registre des procès-verbaux. Séance du 21 Juillet 1864 ", dans J. Boucher de Perthes (éd.), *Antiquités Celtiques et Antédiluviennes. Mémoire sur l'Industrie Primitive et les Arts à leur Origine*, vol. 3, Paris, 1864, 265-268, esp. 265.

45. *Idem*, 266.

46. *Ibidem*.

ters long. The bone was worn and covered by a tightly adhering matrix of the same nature as the surrounding terrain. The excavation proceeded for a long time without anything else being found until Mr. Trancart found part of a human femur or humerus[47]. Some minutes later Trancart recovered a broken portion of a human tibia.

The commission then moved to the second site, about 11 meters away. It is movements like these that remove suspicions the bones were being planted. Dubois[48] stated : " Here again we had to clear away the base of the section to reveal the actual wall of the quarry. The same precautions were taken to assure the homogeneity of the bed and the absence of any disturbance ". At this site, Marcotte found a piece of a human femur, about 13 centimeters long[49]. It came from the bed of yellow brown sand which lies directly on the chalk. Boucher de Perthes[50] noted that two hand axes were also found on the same day.

On July 16, the members of the commission that carried out the July 9 excavation were joined at Moulin Quignon by Mr. Buteux and Mr. de Mercey, members of the Geological Society of France ; Baron de Varicourt, chamberlain of His Majesty the King of Bavaria ; Mr. de Villepoix, member of the Société d'Emulation ; and Mr. Girot, professor of physics and natural history at the College of Abbeville. In addition to the members of the formal commission a dozen other learned gentlemen, including Boucher de Perthes, were present for the new excavations.

Dubois noted in his report that the quarry wall at the chosen spot was undisturbed and without fissures. About the workers, Dubois[51] stated, " Needless to say, during the entire duration of the work, they were the object of continuous surveillance by various members of the commission ". In examining a large chunk of sediment detached by a pick, the commission members found a piece of a human cranium, comprising a large part of the frontal with a small part of the parietal[52]. It was found at a depth of 3.3 meters in the yellow-brown bed that lies just above the chalk[53].

Dubois's report[54] stated : " Immediately afterwards, one of the workers was ordered to attack the same bank at the same height, but 3 meters further to the

47. J. Dubois, " Untitled report of excavation at Moulin Quignon, on July 9, 1864... ", *op. cit.*, 267.

48. *Ibidem.*

49. *Idem*, 268.

50. J. Boucher de Perthes, " Nouvelles Découvertes d'Os Humains dans le Diluvium,... ", *op. cit.*, 237.

51. J. Dubois, " Untitled report of excavation made at Moulin Quignon, on July 16, 1864. Société Impériale d'Emulation. Extrait du registre des procès-verbaux. Séance du 21 Juillet 1864 ", dans J. Boucher de Perthes (éd.), *Antiquités Celtiques et Antédiluviennes. Mémoire sur l'Industrie Primitive et les Arts à leur Origine*, vol. 3, Paris, 1864, 269-272, esp. 270.

52. *Ibidem.*

53. *Idem*, 271.

54. *Ibidem.*

left. The other worker continued to dig at the extreme right. Is it necessary to repeat that all necessary precautions were taken to establish the integrity of the bed there and that the two workers each continued to be the object of scrupulous surveillance ? We went a long time without finding anything resembling a bone. The excavation on the far right side yielded no results whatsoever. Finally, after about three and a half hours, there came to light the end of a bone, of medium size, situated horizontally in the bed. After its exact position was confirmed, Mr. Marcotte himself took from the sand a complete bone, about 13 centimeters long... It was the right clavicle of an adult subject of small size... Measurements showed it was lying 3 meters from the surface, and 2.3 meters horizontally from our starting point ".

Further excavation caused a landslide. The debris was cleared away, however, and the excavation proceeded, yielding a human metatarsal. Several members of the commission, including the geologist Buteux, saw it in place. It was found at a depth of 3.3 meters just above the chalk in the yellow-brown bed. It was situated about 4 meters horizontally from the line where the excavation started[55]. According to Boucher de Perthes[56] the bones from this excavation, and apparently others, were deposited to the Abbeville museum.

I find the account of this excavation extraordinary for several reasons. First of all, it was conducted by qualified observers, including geologists capable of judging the undisturbed nature of the beds. Second, a skilled anatomist was present to identify the bones as human. Third, it is apparent that the workers were carefully supervised. Fourth, some of the human bone fragments were found at points 3 to 4 meters horizontally from the starting point of the excavation and depths over 3 meters from the surface. This appears to rule out fraudulent introduction. Fifth, the condition of the bones (fragmented, worn, impregnated with the matrix) is consistent with their being genuine fossils. I do not see how such discoveries can be easily dismissed.

Summarizing his discoveries, Boucher de Perthes[57] stated : " The osseous remains collected in the diverse excavations I made in 1863 and 1864 at Moulin Quignon, over an area of about 40 meters of undisturbed terrain without any infiltration, fissure, or channel, have today reached two hundred in number. Among them are some animal bones, which are being examined[58].

Among the human remains, one most frequently encounters pieces of femur, tibia, humerus, and especially crania, as well as teeth, some whole and some broken. The teeth represent all ages — they are from infants of two or

55. J. Dubois, " Untitled report of excavation made at Moulin Quignon, on July 16, 1864... ", *op. cit.*, 272.

56. J. Boucher de Perthes, " Nouvelles Découvertes d'Os Humains dans le Diluvium,... ", *op. cit.*, 238.

57. *Ibidem.*

58. *Idem*, 238-239.

three years, adolescents, adults, and the aged. I have collected, *in situ*, a dozen, some whole, some broken, and more in passing through a screen the sand and gravel take from the trenches[59].

Doubtlessly, a lot has been lost. I got some proof of this last month when I opened a mass of sand and gravel taken from a bank long ago and kept in reserve. I found fragments of bone and teeth, which still bear traces of their matrix and are therefore of an origin beyond doubt "[60].

Armand de Quatrefages, a prominent French anthropologist, made a report on Boucher de Perthes's later discoveries at Moulin Quignon to the French Academy of Sciences. Here are some extracts from the report[61] : " In these new investigations, Boucher de Perthes has employed only a very few workers. In the majority of cases, he himself has descended into the excavation and with his own hands has broken apart and crumbled the large pieces of gravel or sand detached by the picks of the workers. In this manner, he has procured a great number of specimens, some of them very important. We can understand that this way of doing things guarantees the authenticity of the discoveries.

On hearing the first results of this research, I encouraged Boucher de Perthes to persevere, and to personally take every necessary precaution to prevent any kind of fraud and remove any doubts about the stratigraphic position of the discoveries...

As the discoveries continued, Boucher de Perthes sent to me, on June 8, 1864, a box containing several fragments of bones from human skeletons of different ages. I noted : 16-17 teeth from first and second dentitions, several cranial fragments, including a portion of an adult occipital and the squamous portion of a juvenile temporal ; pieces of arm and leg bones, some retaining their articulator ends ; pieces of vertebrae and of the sacrum. The specimens were accompanied by a detailed memoir reporting the circumstances of their discovery.

I examined these bones with M. Lartet. We ascertained that most of them presented very nicely the particular characteristics that were so greatly insisted upon in denying the authenticity of the Moulin Quignon jaw. In accord with M. Lartet, I felt it advisable to persuade M. Boucher de Perthes to make further excavations, but this time in the presence of witnesses whose testimony could not in the least be doubted... Among the more important specimens found in these latest excavations are an almost complete lower jaw and a cranium.

59. J. Boucher de Perthes, " Nouvelles Découvertes d'Os Humains dans le Diluvium,... ", *op. cit.*, 240.

60. *Idem*, 241.

61. A. De Quatrefages, " Nouveaux ossements humains découverts par M. Boucher de Perthes à Moulin-Quignon ", *Comptes Rendus Hebdomadaires de l'Académie des Sciences*, 59 (1864), 107-111.

All of these finds were made in the course of excavations that were mounted in an on-and-off fashion, without any definite pattern. That is to say, Boucher de Perthes would suddenly proceed to the sites, sometimes alone and sometimes with friends. Doing things like this very clearly renders any kind of fraud quite difficult. During the course of an entire year and more, the perpetrator of the fraud would have had to go and conceal each day the fragments of bone destined to be found by those he was attempting to deceive. It is hardly credible that anyone would adopt such means to attain such an unworthy goal or that his activities would have remained for so long undetected.

Examination of the bones does not allow us to retain the least doubt as to their authenticity. The matrix encrusting the bones is of exactly the same material as the beds in which they were found, a circumstance that would pose a serious difficulty for the perpetrators of the daily frauds... Because of the precautions taken by Boucher de Perthes and the testimony given by several gentlemen who were long disinclined to admit the reality of these discoveries, I believe it necessary to conclude that the new bones discovered at Moulin Quignon are authentic, as is the original jaw, and that all are contemporary with the beds where Boucher de Perthes and his honorable associates found them ".

I am inclined to agree with De Quatrefages that the later discoveries of Boucher de Perthes tend to confirm the authenticity of the original Moulin Quignon jaw.

At this point, I wish to draw attention to a report by Dr. K.P. Oakley on the Moulin Quignon fossils. It is one of the few scientific reports from the 20th century giving any attention at all to the later discoveries of Boucher de Perthes. Oakley gave the following results from fluorine content testing[62]. The original Moulin Quignon jaw had 0.12 percent fluorine, a second jaw (the one apparently found on June 10) had a fluorine content of 0.05 percent. By comparison, a tooth of *Paleoloxodon* (an extinct elephantlike mammal) from Moulin Quignon had a fluorine content of 1.7 percent, whereas a human skull from a Neolithic site at Champs-de-Mars had a fluorine content of 0.05 percent. Fluorine, present in ground water, accumulates in fossil bones over time. Superficially, it would thus appear that the Moulin Quignon jaw bones, with less fluorine than the *Paleoloxodon* tooth, are recent.

But such comparisons are problematic. We must take into consideration the possibility that much of a fossil bone's present fluorine content could have accumulated during the creature's lifetime. It is entirely to be expected that the tooth of an animal such as an elephant might acquire a considerable amount of fluorine from drinking water and constantly chewing vegetable matter — much more fluorine than the bone in a human jaw, not directly exposed to water and food. Also, the amount of fluorine in ground water can vary from site to site,

---

62. K.P. Oakley, " Relative dating of fossil hominids of Europe ", *Bulletin of the British Museum of Natural History (Geology)*, 34 (1980), 33.

and even at the same site bones can absorb varying amounts of fluorine according to the permeability of the surrounding matrix and other factors. Furthermore, fluorine content varies even in a single bone sample. In a typical case[63],
a measurement taken from the surface of a bone yielded a fluorine content of
0.6 percent whereas a measurement taken at 8 millimeters from the surface of
the same bone yielded a fluorine content of just 0.1 percent. As such, Oakley's
fluorine content test results cannot be taken as conclusive proof that the Moulin
Quignon jaws were " intrusive in the deposits "[64].

If the Moulin Quignon human fossils of Abbeville are genuine, how old are
they ? Abbeville is still considered important for the stone tool industries discovered by Boucher de Perthes. In a recent synoptic table of European Pleistocene sites, Carbonell and Rodriguez[65] put Abbeville at around 430,000 years,
and I think we can take that as a current consensus.

Fossil evidence for the presence of anatomically modern humans at Abbeville is relevant to one of the latest archaeological finds in Europe. Just this year
Thieme[66] reported fading advanced wooden throwing spears in German coal
deposits at Schöningen Germany. Thieme gave these spears an age of 400,000
years. The oldest throwing spear previously discovered was just 125,000 years
old[67].

The spears discovered by Thieme are therefore quite revolutionary. They are
causing archaeologists to upgrade the *cultural* level of the Middle Pleistocene
inhabitants of Europe, usually characterised as ancestors of anatomically modern humans, to a level previously associated exclusively with anatomically
modern humans. Alternatively, we could upgrade the *anatomical* level of the
Middle Pleistocene inhabitants of northern Europe to the level of modern
humans. The skeletal remains from Moulin Quignon, at least some of which
appear to be anatomically modern, would allow this. They are roughly contemporary with the Schöningen spears. Unfortunately, not many current workers in
archaeology are aware of the Moulin Quignon discoveries, and if they are
aware of them, they are likely to know of them only from very brief (and mis-
leading) negative evaluations.

Why have historians and scientists alike been so sceptical of the Moulin
Quignon finds ? I suspect it has a lot do to with preconceptions about the kind
of hominid that should be existing in the European Middle Pleistocene. The
following passage from Trinkaus and Shipman[68] is revealing : " That any

63. M.J. Aitken, S*cience-based Dating in Archaeology*, London, 1990, 219.

64. K.P. Oakley, " Relative dating of fossil hominids of Europe ", *op. cit.*, 33.

65. E. Carbonell, X.P. Rodriguez, " Early Middle Pleistocene deposits and artefacts in the Gran
Dolina site (TD4) of the " Sierra de Atapuerca " (Burgos, Spain) ", *Journal of Human Evolution*,
26 (1994), 291-311, esp. 306.

66. H. Thieme, " Lower Palaeolithic hunting spears from Germany ", *Nature*, 385 (February
27, 1997), 807-810, esp. 807.

67. *Idem*, 810.

68. E. Trinkaus, P. Shipman, *The Neanderthals*, *op. cit.*, 97.

knowledgeable scientist should take the Moulin Quignon jaw seriously as a human fossil appears difficult to fathom in retrospect. Yet, despite the support for the Neander Tal fossils as an archaic, prehistoric human, few knew what to expect. Clearly, many ... still expected human fossils to look just like modern humans ; it was only a matter of finding the specimen in the appropriately pre-historic context ".

It is clear that Trinkaus and Shipman would expect to find only ancestors of the modern human type in the European Middle Pleistocene. And today it would be hard to find a " knowledgeable scientist " who did not share this expectation. It is clear to me, however, that this fixed expectation may have obscured correct apprehension of the human fossil record in Europe and else-where. So perhaps it is good for researchers with different expectations to look over, from time to time, the history of archaeology.

My own expectations are conditioned by my committed study of the San-skrit historical writings of Vedic India (the Puranas), which contain accounts of extreme human antiquity. In his review of *Forbidden Archaeology*, Murray[69] wrote : " For the practising quaternary archaeologist current accounts of human evolution are, at root, simply that. The " dominant paradigm " has changed and is changing, and practitioners openly debate issues which go right to the conceptual core of the discipline. Whether the Vedas have a role to play in this is up to the individual scientists concerned ".

I am hopeful that some individual scientists will in fact decide that the Vedas do have a role to play in changing the conceptual core of studies in human origins and antiquity.

But let us return to the more limited question before us. As far as the finds of human bones at Moulin Quignon are concerned, I would be satisfied if a professor of archaeology at a European university, perhaps in France and Bel-gium, would assign some graduate students to reopen the investigation.

69. T. Murray, " Review of *Forbidden Archaeology* ", *British Journal for the History of Sci-ence*, 28 (1995), 377-379, esp. 379.

Los fundamentos científicos de la red sismológica Mexicana. La participación de las comunidades científicas en la construcción de la red, 1880-1910

Rebeca de GORTARI RABIELA

ANTECEDENTES

Los progresos de la sismología a fines del XIX y principios del XX tomaron varios caminos. A partir de la década de 1880, se modifican sus métodos y se sustituyen en gran medida las hipótesis y teorías por una orientación experimental, con el auxilio de instrumentos cada vez mas precisos. Se buscaba incrementar su medición a través de observaciones sistemáticas, metódicas y continuas y al mismo tiempo avanzar en la comprensión de los fenómenos sísmicos y por ende en la sismología como disciplina científica.

Países como Japón e Italia fueron pioneros en los estudios sísmicos y en el perfeccionamiento de instrumentos, los cuales se extendieron posteriormente a Alemania, Hungría y Estados Unidos.

Otro paso importante fueron los distintos esfuerzos por organizar y sistematizar los estudios sismológicos donde Italia llevo la avanzada a través de la creación de un servicio sismológico. Posteriormente de Alemania saldría la iniciativa para conjuntar esfuerzos a nivel internacional que llevaron a la creación primero de la Asociación Internacional de Sismología con una nutrida participación internacional, figurando entre los países, México y posteriormente de una red de estaciones sismológicas a nivel internacional para contribuir al estudio de la física del Globo.

México participó y contribuyó a distintos niveles al avance de la sismología. En una primera fase, el estudio de los sismos a nivel local estuvo basado fundamentalmente en su registro a través de relatos y escritos que permitieron acumular una gran cantidad de datos de observación que reunidos y analizados comparativamente tuvieron como resultado la formalización de algunas propuestas e interpretaciones sobre el comportamiento y el constante movimiento de la Tierra y en particular de los temblores. En una segunda etapa, con la introducción de los primeros instrumentos, se pudieron hacer observaciones mas precisas y verificables, de las cuales resultaron varios trabajos mas detallados sobre algunos sismos en México. Finalmente, el conjunto de tales esfuerzos y avances sirvieron de justificación para lograr el apoyo de diversas instancias políticas en la planeación y establecimiento de la red sismológica en México, a partir de la cual se pudieron elaborar nuevos estudios[1].

En este trabajo se analiza por una parte la forma en que se conjuntaron tanto a nivel nacional como internacional, los avances en la recolección de datos para la medición sísmica apoyados en los primeros instrumentos y en una organización que aunque incipiente confluyeron en la creación de la red sismológica en México ; y por otra, de que manera la comunidad geológica internacional logró traducir los avances científicos a través de su reconocimiento y presencia a nivel gubernamental y hacerlos coincidir con las necesidades practicas y aun políticas de países como México en los que por su alta sismicidad debían encontrar instrumentos que les permitieran medir estos fenómenos y en ultimo termino controlarlos.

EL INTERÉS POR LOS ESTUDIOS SISMOLÓGICOS

Los estudios geológicos durante el XIX y sobre todo en la segunda mitad ocuparon un lugar importante en la comunidad científica en México. Después del *Ensayo geognóstico sobre la superposición de las rocas en ambos hemisferios* de Alejandro de Humboldt y los *Elementos de Orictognosia* de Andrés Manuel del Río, obras tenidas como clásicas durante la primera mitad del siglo XIX, le siguieron una gran cantidad de investigaciones geológicas a partir sobre todo de la década de 1870. Desde entonces, los estudios sismológicos cobraron relevancia como una de las ramas de la geología aplicada.

Como en otros países, los primeros sismólogos en México fueron ingenieros, geólogos, naturalistas, médicos y todos aquellos personajes " ilustrados " que pertenecían a las sociedades científicas de la época. Así, en un primer momento las primeras iniciativas para dedicar esfuerzos al estudio de la sismología partieron de las asociaciones en las cuales existían comisiones en diversas disciplinas. Para fines del XIX los estudios sismológicos pasaron a formar parte de sus prioridades. En sociedades como la Antonio Alzate desde fines de 1889 se estableció una comisión geodinámica para llevar a cabo estudios sobre sismos, teniendo entre sus propósitos : " formar una estadística tanto de los fenómenos pasados como de los venideros proponiendo a la vez los medios

---

1. P. Sánchez, " Estudio de los temblores en Tehuantepec ", *Anales del Ministerio de Fomento de la República Mexicana*, 1898.

mas adecuados para estudiar de la mejor manera posible todo lo concerniente a este ramo de la geología dinámica. Para integrar esta comisión se solicitará ayuda no solo de los naturalistas de profesión sino de todas aquellas personas que *bondadosamente* se dignen cooperar para el adelanto científico de nuestro país "[2].

Entre los trabajos que hay que mencionar esta el primer estudio sistemático de recopilación estadística sobre sismos en México y que abarco los años de 1523 a 1890 de Juan Orozco y Berra, *Efemérides Seismicas Mexicana durante 1888*[3].

De manera paralela aparecen tanto notas traducidas, como transcripciones en el idioma original sobre los avances que va teniendo la sismología a nivel internacional, esto es a nivel teórico, instrumental y organizativo que dan cuenta del constante intercambio de información que se mantiene con los principales científicos de tal disciplina. Cabe mencionar el caso del Profesor M. Forel, director del Observatorio Telúrico de Berna y encargado de la Comisión Sismológica de Suiza y a quien se debe junto con el italiano M. Rossi una de las escalas para la medición de la intensidad sísmica.

A raíz de estos intercambios, también se obtienen los avances a nivel instrumental, de ahí que son publicados artículos del Prof. ingles John Milne dentro de los cuales puede mencionarse *Modern forms of pendulum seismometers (their development and tests)* del cual se basaron algunos de los adelantos mas importantes en instrumentas sismométricos.

De otra parte aparecen iniciativas sobre la necesidad de establecer los principios de una organización para el levantamiento de los registros sísmicos.

Un aspecto que llama la atención es que la interacción que mantienen con el exterior va mas allá del intercambio de publicaciones, ya que incluso en estos años se tiene noticia de la visita de Orozco y Berra al Observatorio Geodinámico Central de Roma, dirigido por el Prof. Miguel E. Rossi[4].

Finalmente es de señalarse que autores como F. Montessus de Ballore que aparece como socio corresponsal de la *Revista de la Sociedad Científica Alzate* en París, mantuvo un constante contacto con México, como lo muestra la publicación de su trabajo *México Seismico* publicado en 1892, donde muchos de los datos podrían haber sido provistos por miembros de la sociedad[5].

2. *Revista mensual de la Sociedad Científica Antonio Alzate, 1889-1890*, t. IV, num. 9 (1890).

3. J. Orozco y Berra, " Efemérides seísmicas mexicanas ", *Memorias de la Sociedad Científica Antonio Alzate*, 1888 ; J. Orozco y Berra, " Adiciones y rectificaciones a las efemérides seísmicas mexicanas ", *Memorias de la Sociedad Científica Antonio Alzate*, 1888.

4. " Memorias de la Sociedad Científica Antonio Alzate ", *Revista mensual científica y bibliográfica*, núm. 6 (diciembre 1888) ; " Memorias de la Sociedad Científica Antonio Alzate ", *Revista mensual científica y bibliográfica*, t. III (1889).

5. F. de Montessus de Ballore, " México seísmico ", *Memorias de la Sociedad Científica Antonio Alzate*, 1892.

LA ORGANIZACIÓN DE LOS PRIMEROS LEVANTAMIENTOS SÍSMICOS

A raíz de la iniciativa de establecimiento de la Comisión Geodinámica, la Sociedad Alzate envío una circular a sus corresponsales en diversas partes del país con el propósito de solicitarles tanto el envío de todas aquellas noticias que pudieran servir para formar una estadística de la historia sísmica de México, como también de proponerles la manera en como levantar los datos sísmicos. Al respecto hay que señalar que esta se apoyaba en los avances en la teoría sismológica que hasta ese momento explicaban las causas y tipos de temblores, así como la manera de observarlos a partir del carácter, dirección, intensidad y duración del movimiento y que daban cuenta de los primeros intentos por unificar criterios para una medición sísmica mas confiable que la obtenida a través de los testimonios de la gente.

Los avances que se lograron no hubiesen sido suficientes sin la vinculación de esta comunidad con el gobierno mexicano, el cual se encargaría de apoyar y financier estos esfuerzos, como parte de su proyecto modernizador, donde el interés por desarrollar disciplinas como la sismología revestía varios ángulos.

Así, las observaciones sismológicas empezaron a ser incluidas como parte de los trabajos de las diversas instancias creadas para el desarrollo e impulso de las exploraciones geológicas y mineras. De esta manera, la Comisión Geográfico Exploradora creada en 1877, integrada por un gran numero de personajes ilustrados de la época, realizaron también estudios sismológicos.

En algunas comisiones locales, desde sus inicios quedo establecido como parte de sus funciones el levantamiento de datos sismológicos como lo muestran las *Instrucciones para el estudio de los fenómenos sísmicos* establecidas en 1887 por el Ing. Agustín Díaz, director de la Comisión Geográfico-Exploradora y de la Comisión Científica de Sonora : " 1° Descripción ortográfica e hidrográfica de la región explorada ; 2° Estudio geológico de la misma ; 3° Construcción de los perfiles geológicos de los principales caminos recorridos ; 4° Construcción de un croquis topográfico para situar en él las grietas originadas por el temblor ; 5° Bosquejo geológico de la región explorada ; 6° Levantamiento del plano de las ruinas del pueblo de Babispa ; 7° Formación de una colección de rocas para su estudio micrográfico y clasificación precisa, indicando su composición mineralógica, acompañada de observaciones acerca del papel que han desempeñado en el levantamiento de las cordilleras ; 8° Estudio y discusión de los fenómenos originados por el temblor del 3 de mayo de 1887, tales como dislocaciones, hundimientos, agrietamientos, derrumbes, inundaciones, emisión de gases en ignición, etc. ; 9° Marcha, duración y manera de propagación de los movimientos sísmicos ; 10° Determinación del epicentro o foco superficial del temblor, y profundidad aproximada del foco real ; 11° Construcción de una carta con las curvas isosísmicas de igual intensidad del temblor, y si fuere posible construcción las curvas cosísmicas ; 12° Causa probable del temblor "[6].

Estas instrucciones se redactaron al año siguiente del temblor ocurrido en 1887 en Babispa, Sonora que de acuerdo a estudios recientes fue uno de los mas grandes del siglo pasado en México : También es de mencionarse la Comisión Geológica de México creada en 1886, que contó con un gabinete especializado para la preparación de mapas, cartas, secciones transversales, columnas de cortes geológicos, etc. Finalmente, a nivel de toma de decisiones, la participación que tuvo esta comunidad en la conducción y establecimiento de políticas quedaría demostrada con la creación del Instituto Geológico de México establecido en 1891, y que se dedico a hacer investigaciones de geología y de varias ramas de la geología aplicada entre las que estaban los estudios sismológicos. Su dependencia de la Secretaría de Fomento, le permitió contar desde entonces con la subvención del gobierno, convirtiéndose además en el vocero de la comunidad geológica tanto internamente como a nivel internacional.

PROPUESTAS PARA LA EXPLICACIÓN DE LOS SISMOS Y AVANCE
DE LA SISMOLOGÍA

Los trabajos que se elaboraron a partir de la década de 1880 en México tuvieron dos vertientes, una interpretativa donde cabían las mas diversas hipótesis y teorías sobre las causas de los temblores, y otra apoyada en un cumulo de datos recogidos por distintas vías que aunque no eran sistemáticos y confiables han servido como herramientas para el estudio histórico de los sismos. A estos trabajos se incorporarían posteriormente los elaborados con los datos suministrados por los sismógrafos.

Entre los primeros esta un estudio aparecido en el número 9 de la *Revista de la Sociedad Científica Antonio Alzate* publicado en marzo de 1890 : " En trabajos anteriores publicados por la Sociedad Alzate se ha mostrado por medio de unas 50.000 sacudidas acaecidas en todas partes del mundo, que no tienen relación ninguna con las horas del día, las culminaciones de la luna y las estaciones astronómicas. De todas las leyes enunciadas sobre los temblores, estas son las mas comúnmente aceptadas por los sismólogos. Todo esto no es verdadera ciencia, deben buscarse las causas de los sismos en donde se producen que no es en el exterior, sino en el interior de la corteza terrestre, presentándose justamente en esta ".

No obstante lo temprano de la afirmación anterior, varios autores aun cuando reconocían que la mayoría de los temblores tenía un origen tectónico, consideraba que sus causas podrían originarse en corrientes magnéticas producidas por el paso de meteoros como afirmaba Miranda y Marrón en su artículo

---

6. J.G. Aguilera, " Estudio de los fenómenos seísmicos del 3 de mayo de 1887 ", *Anales del Ministerio de Fomento de la República Mexicana*, 1888.

sobre *Los terremotos del año 1908*[7]. Otra discusión que resalta y que en México fue objeto de gran atención debido a su constitución volcánica, fueron los trabajos referidos a la existencia o no de relaciones entre los sismos y los volcanes. Para autores como Emilio Bose[8], la distinción entre los temblores volcánicos y los tectónicos solamente era posible si se utilizaban instrumentas registradores.

Años mas tarde, en el trabajo de Paul Waitz y Fernando Urbina sobre *Los temblores de Guadalajara en 1912*[9], se afirmaba que : " Los temblores de origen tectónico se verifican igualmente en regiones enteramente cubiertas por focos volcánicos. Esta conclusión viene a desvanecer el criterio que se tenia anteriormente de que en las regiones volcánicas, los temblores eran producidos por esos focos, desechando toda intervención tectónica. En esta región, Acambay y Guadalajara, la actividad sísmica fue enteramente independiente de la actividad volcánica, pues no se encontró la menor huella de esta ultima, no obstante que son regiones cubiertas enteramente por aparatos eruptivos, algunos de los cuales se han formado en épocas modernas ". Posteriormente en 1926 Muñoz Lumbier, en la *Memoria descriptive de la carta sísmica de México*[10] asentaba que : " No hay razon científica alguna para fundar la opinión de que el volcanismo haya intervenido en estos temblores, y si hay, por el contrario datos de observación suficientes, para asegurar que dichos movimientos han sido de origen tectónico ".

Un tercer tipo de trabajos que proliferaron desde fines de 1880 son los estudios detallados sobre diversas regiones del país, realizados por miembros de la Comisión Geográfico-Exploradora y de las Comisiones Científicas, las cuales procuraron que a partir de entonces hubiera cierta sistematización y uniformidad en los datos recogidos. Un ejemplo es el trabajo de José G. Aguilera, denominado " Estudio de los fenómenos seismicos del 3 de mayo de 1867 ", en el que el autor señalaba que para medir el sismo ubicado en Sonora había utilizado la metodología de Dutton y Hayden de la Comisión Geológica Norteamericana. Entre los estudios mas detallados están los de Emilio Bose[11], quien junto con otros ingenieros realizaron el estudio de los temblores in situ, en

7. M. Miranda y Marrón, " Los terremotos del año del 1908 ", *Memorias de la Sociedad Científica Antonio Alzate*, t. 28 (1909-1910) ; M. Miranda y Marrón, " Las catástrofes de 1906 ", *Boletín de la Sociedad Mexicana de Geografía y Estadística de la República Mexicana*, 1907.

8. E. Bose, " Sobre las regiones de temblores en México ", *Memorias de la Sociedad Científica Antonio Alzate*, 1902.

9. P. Waitz y F. Urbina, " Los temblores de Guadalajara en 1912 ", *Boletín del Instituto Geológico de México*, 1919.

10. M. Muñoz Lumbier, " Memoria descriptiva de la carta sísmica de México ", *Boletín del Instituto Geológico de México*, 1926.

11. E. Bose, " Informe sobre los temblores de Zanacatepec : a fines de septiembre de 1902 ; y sobre el estado actual del Volcán de Tacaná ", *Parergones del Instituto Geológico de México*, 1903 ; E. Bose y E. Angermann, " Informe sobre el temblor del 16 de enero de 1902 en el Estado de Guerrero ", *Parergones del Instituto Geológico de México*, 1904 ; E. Bose, *et al.*, " El temblor del 14 de abril de 1907 ", *Parergones del Instituto Geológico de México*, 1908.

donde recogieron información a partir de entrevistas con los habitantes donde ocurrió el temblor, la cual era contrastada con registros sismológicos de otros países y con la metodología de medición de la frecuencia sísmica de Forel/ Rossi.

A partir de estas primeras mediciones se pudo avanzar en la unificación de los criterios para la construcción de la sismología como disciplina científica.

AVANCES EN LA PARTE ESTADÍSTICA Y DE ORGANIZACIÓN DE LAS OBSERVACIONES

Junto a las aportaciones al conocimiento e interpretación sismológica, de manera paralela, en México, también se avanzo en la parte organizativa y de sistematización. Desde 1880 como señalábamos anteriormente uno de los principales promotores fue Juan Orozco y Berra, miembro de la Sociedad Alzate, quien señalaba :

" Es un hecho reconocido … que la superficie de nuestro planeta esta sujeta a continuas movimientos, muchos de ellos han escapado a nuestra observación y escaparán aún si no existen delicados y sensibles aparatos que revelen esas alteraciones no son únicamente los grandes movimientos más o menos intensos los que se observan, no la seismología, esa nueva rama de la geología dinámica va más adelante y trata de estudiar los movimientos pequeñisimos que por tanto tiempo escapan a los sentidos del hombre y a los que se ha dado la denominación de microseismos… Nosotros no tenemos ni tiempo ni elementos intelectuales ni materiales para emprenderlo, y nos contentamos con hacer un llamamiento a las personas de buena voluntad que, contribuyendo al estudio geodinámico, puede obtener datos precisos para la ciencia y honra y provecho para el país y para si …deben verificarse numerosisimos sacudimientos seismicos, cuyo estudio científico y estadístico sería de gran importancia para el adelanto de la ciencia, y hoy dejamos perder en su mayor parte estérilmente, pues solo aprovechamos los más notables. Lo poco que se hace en el país sobre movimientos microséismicos se debe a la iniciativa de un particular el Sr. Carlos Molt, socio corresponsal de Orizaba, que con una constancia y actividad dignas de gran encomio ha establecido a sus expensas, un pequeño observatorio, ¿ cuando entrará México en honrosa competencia con el Japón, Suiza, Italia ? ".

Este llamado a conjuntar esfuerzos para recabar los datos y contar con los instrumentos necesarios para registrar con mayor fidelidad los movimientos sísmicos fue recogido por las asociaciones científicas locales, además de la adquisición de sismógrafos para las estaciones meteorológicas establecidas a iniciativa de la Secretaría de Fomento en 1894.

En el caso de las estaciones meteorológicas prácticamente desde su establecimiento, les fueron giradas instrucciones para registrar los sismos, ruidos subterráneos y erupciones volcánicas ; fenómenos que debían comunicarse por vía telegráfica a la Dirección del Servicio Meteorológico. Para ese entonces, se tenia la idea de dotar de sismógrafos a las diferentes Estaciones Meteorológicas[12], con aparatos del mismo modelo y de las mismas dimensiones, cuyos trazos fueran fácilmente comparables y dieran idea de la manera como se propaga el fenómeno. El modelo que se pensaba adoptar " por ser el mejor, es el que el inteligente mecánico Don Ramón Alba presentó en las Exposiciones Universales de París y de Buffalo, pero más pequeño para que su costo sea más módico ".

Los aparatos con los que se dotó a los observatorios meteorológicos de acuerdo con un autor de la época eran muy defectuosos y en realidad es hasta 1904 cuando en el Observatorio Astronómico de Tacubaya se instalaron dos sismógrafos Omori, cuando se pudo iniciar el estudio racional de los sismos[13]. Un ejemplo de iniciativas individuales son los esfuerzos del presbítero Severo Díaz quien además publico varios trabajos[14].

El establecimiento de los primeros aparatos ya no de manera aislada sino con el propósito de montar un sistema aunque incipiente, estuvo marcado por un primer intento de sistematización en el levantamiento de los datos, que después seria retomado.

EL ESFUERZO NACIONAL UNIDO A LA ORGANIZACIÓN CIENTÍFICA

INTERNACIONAL

Desde la década de 1870 existían instituciones científicas cuya misión fue centralizar en Nueva Zelanda, Filipinas y las Indias Holandesas todas las observaciones realizadas correspondientes a los terremotos sentidos en esas regiones. Otro esfuerzo fue el emprendido por Rossi que en 1873 reunía en su *Bulletino del Vulcanismo Italiano* todas las observaciones de carácter sísmico realizadas por las estaciones italianas. Años mas tarde en 1879, Tachini transformó la organización en un servicio del Estado que termino por ser la Ufficio Centrale di Meteorologia e di Geodinámica. El mismo año se estableció la Seismological Society of Japan, la cual gracias a John Milne seguido de su discípulo Fusakushi Omori puso a Japón a la cabeza de las naciones interesadas en esta ciencia. Esta institución tuvo carácter oficial en 1892 y su publicación *Earthquake Investigation Committee* recoge los trabajos más importantes referentes a la sismología mundial.

---

12. M. Pastrana, *Instrucciones para las estaciones meteorológicas del Servicio Meteorológico de la República Mexicana,* México, 1904.

13. *Primer Centenario de la Sociedad Mexicana de Geografía y Estadística, 1833-1933,* t. 1, México, 1933 ; " El Observatorio Meteorológico de León ", *Revista Científica y Bibliográfica de la Sociedad Científica Antonio Alzate,* 1894.

14. S. Díaz, " Estudio sobre los temblores sentidos en Guadalajara : en el año de 1912 ", *Observatorio Meteorológico y Astronómico del Seminario,* 1912.

En 1889 Von Rebeur Paschwitz descubrió que las ondas sísmicas debidas a los terremotos pueden ser registradas en todo el mundo por aparatos de suficiente sensibilidad, lo que le dió la idea de crear una asociación internacional que cristalizó en la Asociación Internacional de Sismología. A partir de entonces se propagó por otros países, gracias al propio Paschwitz, al maestro de Gotinga, Emilio Wiechert y a los profesores Gerland, Hecker, Rudolph y Mainka. Como expresion de estos avances aparecen varias publicaciones : *Beitrag für Geophysik* (Leipzig) : *Bulletino della Societá Sismológica Italiana* (Modena) ; *Bulletin of the Imperial Earthquake Investigation Committee* (Tokio) ; *Publications du Bureau Central Sismologique International* (Estrasburgo) ; *The International Seismological Summary-Formerly ; The Bulletin of the British Association Seismology Committee*, continuación de la *British Association for the Advancement of Science*, entre otras.

Si bien en México en sus inicios los instrumentas sismológicos se encontraban instalados al lado de los meteorológicos, en general la investigación sismológica con el establecimiento del Instituto Geológico de México se puede considerar que dió inicio su institucionalización. A partir de entonces, este se encargo de los estudios sismológicos del país y de su difusión a través de publicaciones especiales, artículos en la prensa periódica y boletines de sociedades científicas. El Instituto también desde su fundación ensanchó considerablemente sus relaciones y comunicación con casi todas las académicas científicas tanto del país como a nivel internacional a través del intercambio de registros, estudios, información sobre métodos de medición y de instrumentación, así como de su activa participación en los eventos relacionados con la sismología. Como ejemplo, en 1901 acudió a la primera Exposición de Instrumentos Sísmicos efectuada en ocasión de la Primera Exposición Universal, donde fue presentado el sismógrafo mencionado anteriormente. También desde finales del siglo XIX participó en varios congresos, entre ellos el de la Unión Geodésica y organizó en 1906 en México la XX sesión del Congreso Internacional de Geología. Posteriormente, el Instituto asistió como delegado representante del país a las conferencias de la Asociación Internacional de Sismología.

Desde fines de 1890 sismólogos como Milne, Rebeur-Paschwitz, Ehlert y Gerland habían hecho constantes esfuerzos por establecer una organización encargada de los estudios sismológicos. Así hicieron llamamientos en los Congresos Internacionales de Geografía reunidos en 1895 en Londres y en 1899 en Berlín para estimular la fundación de una asociación internacional. Desde ese año se nombro una comisión permanente para los estudios de sismología, pero no se llego al establecimiento de la organización. Quien trabajo mas en ese sentido fue el profesor de Geografía de la Universidad de Estrasburgo G. Gerland ; quien logró que el gobierno alemán fundara una Estación Central Sismológica para Alemania, pero con las dimensiones y la dotación suficiente para poder servir de oficina central a una organización internacional de estudios sobre temblores, la cual fue construida entre 1899 y 1900. Desde entonces dirigió parte de sus esfuerzos a crear una Asociación semejante a la Geodésica Internacional, donde el gobierno alemán invitara a las otras naciones a enviar a sus delegados a una conferencia en Estrasburgo para fundar la citada asociación[15].

Finalmente en abril de 1902 logró reunir a los delegados de Austria, Hungría, Italia, Dinamarca, Bélgica, Japón, Rusia, Sajonia y Suiza, quienes discutieron las bases de una asociación internacional proponiéndose para ello elaborar los estatutos de una asociación de Estados que dejaría a cada país en libertad para la organización de su servicio sismológico.

La Segunda Conferencia se celebró también en Estrasburgo en julio de 1903 con una mayor participación, asi acudieron : Suiza, Rusia, Rumania, Portugal, Holanda, Japón, Italia, Hungría, Austria, Inglaterra, EU, España, Congo, Chile, Bulgaria y Bélgica. Como invitado del gobierno alemán acudió el Ing. José G. Aguilera, director del Instituto Geológico de México en representación de México.

Dicha presencia fue recogida en una crónica de la reunión en donde se asentaba que " El fin de la asociación creada allí es el fomento de todas las tareas de la sismología, que solo se pueden resolver por la cooperación de numerosas estaciones seismológicas repartidas sobre toda la tierra. Los medios principales para esto son : observaciones según principios comunes ; experimentos para cuestiones especiales particularmente importantes ; fundación y subvención de observatorios seismológicos en paises que necesitan la ayuda de la asociación ; organización de una oficina central para la colección y elaboración de los informes de los diferentes paises… La Oficina Central está conectada con la Estación Central Imperial Seismológica en Estrasburgo de tal manera que el Director de esta es al mismo tiempo Director de la Oficina Central. La Oficina Central recoge los informes de los diferentes países, los ordena en revistas generales y las publica "[16].

Entre 1903 y 1906 tuvieron lugar varias reuniones (Francfort, 1904 y Berlín, 1905) que permitieron la consolidación de la Asociación Sismológica Internacional. A la par que en países como México se organizaba y formalizaba la idea de establecer un Servicio Sismológico Nacional[17].

En la siguiente reunión de 1906 en Roma, Italia, entre los tópicos que se discutieron estaban :

15. " La première conférence sismologique internationale de Strasbourg ", *Revista Científica y Bibliográfica de la Sociedad Científica Antonio Alzate*, 1902.
16. " La organización del estudio de los temblores sobre toda la tierra ", *Revista Científica y Bibliográfica de la Sociedad Científica Antonio Alzate*, 1903.
17. R. Aguilar Santillán, " Apuntes relativos a algunos Observatorios e Institutos Meteorológicos de Europa ", *Boletín de la Sociedad de Geografía y Estadística de la República Mexicana*, 1890.

- El tipo de instrumentos que tendrían que instalarse en las estaciones, argumentándose la conveniencia de que fueran de una sola clase aunque aún no se había decidido por alguno en específico.

- La necesidad de unificar los sismogramas y de concentrarlos y difundirlos en catálogos a nivel internacional.

- Proporcionar indicaciones sobre la manera de medir el tiempo, que de acuerdo a la resolución de las dos primeras conferencias sismológicas, el tiempo seleccionado era el de Greenwich. Se decidió que se elaboraría un cuestionario que seria enviado a los observatorios.

- La necesidad de seleccionar los mejores puntos de observación en la superficie terrestre.

- La organización de las observaciones sísmicas. Forel recordó que hasta ese momento se había estudiado sobre todo la sismografía que es la parte en progreso de la ciencia sismológica. Pero que era necesario referirse a los catálogos sísmicos que son necesarios para estudiar la sismicidad de los diferentes países del mundo y las cuestiones de periodicidad de los sismos, de tal modo que se pudiera contar con una estadística completa.

Para la reunión que tuvo lugar en 1907 en La Haya, México figuraba ya como asociado y al año siguiente en abril de 1908, fue decretada la creación de su Servicio Sismológico Nacional dependiente del Instituto Geológico. Con ello se cerraba esta primera etapa donde se llegó al acuerdo de instalar estaciones en aquellos países donde no existieran para crear una red a nivel internacional y poder contribuir a la construcción de un mapa sismológico a nivel mundial.

Como señalaban en uno de los primeros trabajos apoyados en los registros de las recién creadas estaciones sismológicas en México : " El problema mecánico se pudo estudiar con los datos instrumentales sacados de los seismogramas registrados en las estaciones de la Red Sismológica Mexicana. ...Justo es mencionar este adelanto que, aunque para una gran parte del publico pasa inadvertido, para los que aquí en México se dedican sinceramente a la Sismología, les suministra datos de gran valor ; porque además de obtener de los seismogramas la primera indicación aproximada del lugar mas fuertemente conmovido, sirve de auxilio de la geología para mejorar fórmulas, corrigiendo los coeficientes y adaptándolos a nuestro país, y en general para adquirir conocimientos acerca del modo de ser del movimiento sísmico : velocidad de propagación de ondas, intensidad, amplitud, periodo, etc. "[18].

El avance tan importante y rápido que tuvo la sismología en este periodo puede sintetizarse en las palabras de bienvenida del representante del gobierno

---

18. F. Urbina, H. Camacho, " La zona megasísmica de Acambay-Tixmadeje Estado de México : conmovida el 19 de noviembre de 1912 ", *Boletín del Instituto Geológico de México*, 1913.

italiano a la reunión de la Asociación Internacional de Sismología que tuvo lugar en octubre de 1906 en Roma, Italia[19] : " Ninguna consideración debe prevalecer frente a la importancia de los estudios sobre los temblores que bien que afectan mas seguido y de manera mas sensible a algunas regiones del globo, sin embargo son del interés del conjunto de la superficie y pueden ser estudiados con instrumentos delicados incluso a una distancia muy grande. La asociación de todos los sabios del mundo que recogen y concentran sus observaciones en un instituto único no dejara de dar un impulso mayor, una unidad armónica a los nobles objetivos que se persiguen en estos estudios ".

19. " Estación seísmica del Doctor Timoteo Bertelli en Florencia ", *Memorias de la Sociedad Científica Antonio Alzate*, 1889.

LAS BASES DE LA INSTRUMENTACIÓN SÍSMICA MEXICANA : ANÁLISIS DE LA INTEGRACIÓN DE LA RED SISMOLOGICA MEXICANA 1889-1910

María Josefa SANTOS

ANTECEDENTES

Desde mediados del siglo XIX las comunidades científicas mexicanas habían solicitado al gobierno mexicano en distintos foros primero la adquisición de instrumentos confiables que permitieran uniformar las medidas sísmicas nacionales y después con la creación del Instituto Geológico en 1891, el establecimiento de una red sismológica nacional que otorgara la posibilidad a investigadores nacionales pero también extranjeros de contar con medidas uniformes y confiables de los eventos sísmicos de un país tan " conmovido " como el nuestro. Al respecto don Juan Orozco y Berra escribía en 1888 : " Es un hecho reconocido hoy día, que la superficie de nuestro planeta esta sujeta a continuos movimientos, muchos de ellos han escapado a nuestra observación y escaparán aún si no existen delicados y sensibles aparatos que revelen esas alteraciones. No son únicamente los grandes movimientos más o menos intensos los que se observan, no, la seismología, esa nueva rama de la geología dinámica va más adelante y trata de estudiar los movimientos pequeñisimos que por tanto tiempo escapan a los sentidos del hombre y a los que se ha dado la denominación de microseismos "[1].

Estas primeras peticiones estaban siempre acompañadas de referencias comparativas con los avances de la nueva ciencia en el extranjero, así por ejemplo Orozco y Berra en el mismo documento hace referencia a los trabajos de la comisión sismológica de Suiza que estableció un programa de trabajos encaminado no solamente a conocer los sismos presentes y futuros sino también a hacer una reconstrucción sísmica inmediata a partir de testimonios comparados

---

1. J. Orozco y Berra, " Efemérides sísmicas mexicanas ", *Memorias de la Sociedad Científica " Antonio Alzate "*, 1888, 5.

y confrontados entre los diferentes habitantes y científicos suizos. Otro de los países más envidiados en cuanto al avance y medición de los movimientos sísmicos fue Italia que ya desde 1887 contaba con 678 lugares de observación sísmica en 1887 (donde se tenían desde sismoscopios hasta telégrafos para enviar datos) situados en parajes distantes entre sí por 20 kilómetros establecidos de acuerdo a una red de malla.

Estos primeros discursos dan cuenta de dos de las cuestiones que se pretenden mostrar en este trabajo. La primera el enorme interés de la comunidad científica mexicana por el establecimiento de una red sismológica confiable mismo que se vio recompensado con la compra de instrumentos italianos y japoneses que se establecieron en distintas estaciones meteorológicas y también con el desarrollo de aparatos realizado por los propios científicos mexicanos. La segunda cuestión sería la enorme vinculación de estos primeros sismólogos con comunidades científicas internacionales al grado de que el diseño, ubicación e instrumentación de todas y cada una de las estaciones mexicanas estuvo regido por criterios establecidos por aquellas lo cual eventualmente facilitaría la comparación de los datos registrados. Esto último permitió una carta sísmica nacional que contribuiría en la definición de las regiones de riesgo sísmico a nivel mundial. Para tratar estas dos cuestiones se ha dividido el trabajo en tres partes en donde los intereses nacionales e internacionales se encuentran muy ligados : la primera da cuenta de las primeras estaciones e instrumentos que son los antecedentes inmediatos de la red sismológica nacional mexicana inaugurada el 6 de septiembre de 1910 por el entonces presidente el general Porfirio Díaz como parte de los festejos del primer centenario de la independencia mexicana. La segunda se refiere a los antecedentes inmediatos políticos, científicos y financieros que definieron finalmente las características de la red. Por último, en la tercera parte mencionamos algunos de los discursos y motivos políticos que ayudaron finalmente a convencer al gobierno porfirista de financiar tan costoso proyecto.

PRIMEROS INSTRUMENTOS, PRIMERAS ESTACIONES

Como se ha mencionado brevemente, las primeras estaciones sismológicas estuvieron albergadas en la red nacional de estaciones meteorológicas. Esta situación implicó que en un principio el desarrollo de la sismología estuviera ligado al de la meteorología con la consecuente competencia por recursos de ambas disciplinas.

Por otro lado, antes del establecimiento de los primeros aparatos especializados, en México, se habían utilizado distintos instrumentos para la medición de los sismos, que iban desde las percepciones de los afectados, incluyendo observaciones de ciertos " objetos péndulo " por ejemplo ; lámparas, campanas de la iglesia, etc., hasta instrumentas más sofisticados como por ejemplo una especie de reloj con un dispositivo de escape que se disparaba al producirse el sismo ; de este modo se paraban las manecillas de aquél y quedaba así marcado el momento de la primera sacudida. Para determinar la dirección de ésta se empleaba una vasija aplanada, que contenía mercurio, y estaba provista de cuatro orificios ; la cantidad de mercurio lanzada por uno de estos orificios indicaba la dirección e intensidad de la sacudida.

Mucho más perfectos eran los péndulos construidos por Zöllner y Rebeur Paschwitz, que entraban en oscilación por el movimiento del suelo ; un aparato inscrito de estructura análoga a la de los termógrafos barógrafos registraba en la banda de papel no sólo el principio, sino la duración e intensidad de los estremecimientos del suelo. Estos primeros dispositivos tenían el inconveniente de registrar fielmente sólo el principio del estremecimiento, pero no su continuada serie de movimientos, pues una vez que el péndulo entra en oscilación por la primera sacudida, continua en su movimiento periodico propio y, por lo tanto, no reproducían el estremecimiento real del suelo, sino el decrecimiento gradual de la oscilación debida al primer impulso[2].

Pero volvamos a los primeros recintos que albergaron los primitivos instrumentos que precedieron a la moderna y confiable red mecánica que arrancó con la estación de Tacubaya. Los primeros observatorios datan de 1876 año en el que el Ing. Ángel Anguiano se encargó de la formulación del proyecto y construcción de un observatorio siendo la idea del gobierno que : " aquel lugar se destine definitivamente a un objeto digno y útil, y que corresponde a las exigencias actuales de la ciencia y a nuestra cultura, el proyecto deberá comprender no solamente un observatorio astronómico, sino además un observatorio meteorológico y magnético, dependiente de la Secretaría de Fomento ".

Desde este primer proyecto se pretendía que el observatorio de Chapultepec se encontrara relacionado con otros, que se llamarían de primer orden que a la vez exigirían otros de orden secundario, justo los criterios seguidos después para el establecimiento de la red sismológica. Como en la red sismológica los criterios para establecer los primeros observatorios serían los que difirieran notablemente por su altitud y condiciones climatológicas del Observatorio Central, p.e. Veracruz, Orizaba, México ; los secundarios reconocerían como centro inmediato de sus operaciones al de primer orden de aquel lugar, que con poca diferencia se encontrase bajo las mismas condiciones de clima y de localidad. También en la ciudad de México se pretendían establecer varios observatorios (Tacubaya, Chalco, Texcoco y Zumpango) que formarían un cuerpo de preciosos datos para la ciencia. Los trabajos del observatorio comenzaron hasta el 6 de marzo del año siguiente 1877, y una vez establecido este se procedió a formar un grupo de observadores voluntarios, que sin estipendio alguno, prestarán su valiosa colaboración haciendo observaciones meteorológicas de acuerdo con las instrucciones del observatorio central. Lo mismo sucedió con

2. F. Frech, *Geología I : volcanes, estructura de las montañas, temblores de Tierra*, Barcelona, 1926.

las observaciones sísmicas para las cuales incluso se inventaron algunos instrumentos que ayudarían a precisar los datos obtenidos siguiendo escalas más empíricas como la de Forel Rossi. Entre las justificaciones políticas científicas que pedían la creación del observatorio se encontraba la continua preocupación de sus impulsores por llevar a la meteorología al rango de ciencia exacta cosa que solo se podría lograr a partir de observaciones reiterativas, precisas y " concienzudas ". Es importante esta justificación porque este argumento es uno de los que se esgrimió durante la creación de la red sismológica. Desde su establecimiento el Observatorio Central (localizado en el Palacio Nacional) estuvo conectado a la red telegráfica nacional y daba cuenta a partir de los tres sismógrafos instalados en él de los acontecimientos sísmicos del país. El primero adquirido por el gobierno mexicano en 1894 era un sismoscopio Palmieri construido por la casa Negretti y Zambra de Londres. Los segundos instalados en el observatorio de Tacubaya en 1904 eran dos sismógrafos Omori con los cuales, según relatan fuentes de la época, se inició el estudio racional de los sismos. Fue en este observatorio central cuando a principios de siglo (en 1904) se publicaron las primeras instrucciones a las estaciones meteorológicas para registrar sismos, temblores de tierra, ruidos subterraneos y erupciones volcánicas. Estos fenómenos geológicos deberán comunicarse inmediatamente por telégrafo a la Dirección del Servicio Meteorológico del país diciendo la hora en la que se produjeron.

Llegados a este punto tenemos que rescatar dos cosas, la primera la relevancia de la red telegráfica para la cabal comunicación de todo tipo de datos entre las incipientes comunidades científicas ; y segunda la presencia importante de los colaboradores regionales organizados de tal suerte que cada estado absorbería el presupuesto de sus propias estaciones además de que los encargados de éstas no cobraban honorarios, muy por el contrario pertenecer a estas comunidades científicas significaba un honor difícil de despreciar. Pero continuemos con las instrucciones giradas por los primeros hombres de la ciencia mexicana : " en el caso de temblores deberá decirse también la duración del temblor y si fue trepidatorio, oscilatorio o giratorio o si participó de dos de estos movimiento o de tres. En las estaciones provistas de aparatos especiales llamados sismógrafos, se deberán sacar tres copias del trazo hecho por el instrumento, para archivar una y para remitir las otras dos a una ala de la Dirección del Servicio Meteorológico del país "[3].

Como lo sugiere la nota anterior no todas las estaciones meteorológicas estaban provistas de un sismógrafo, y se esperaba para hacerlo, a contar con aparatos del mismo modelo y de las mismas dimensiones, cuyos trazos fueran fácilmente comparables y ayudaran a definir la forma en que se propaga el fenómeno. Para hacer más precisas las mediciones en las estaciones que no

---

3. P. Sánchez, " Estudio de los temblores en Tehuantepec ", *Anales del Ministerio de Fomento de la República Mexicana* (1898).

contaban con sismógrafos se sugería una forma de implementar uno : " si se quiere, puede improvisarse un sismógrafo muy elemental colgado de un hilo de una viga de un techo sobre el cual no haya habitación, o mejor de una ménsula fijada a un muro y fiando a su extremo inferior una bala provista en el extremo inferior de su diámetro vertical de una aguja muy fina soldada a ella. Debajo de esta bala se pone una superficie horizontal cubierta de arenilla fina sobre la cual trazará la aguja la dirección del movimiento, la que podrá referirse al meridiano astronómico si se ha tenido cuidado de señalar dicha linea en la superficie horizontal de la tierra "[4].

Además, las estaciones meteorológicas contaban con otra clase de instrumentos que si bien no habían sido diseñados para tomar mediciones sísmicas, sí contribuyeron en las primeras mediciones de los temblores. Uno de estos instrumentos sin duda el que desató mayor polémica fue el pantógrafo que sólo o combinado con algún microsismógrafo contribuyo también a medir la longitud de las ondas sísmicas.

Por otro lado, al estar las comunidades científicas mexicanas muy ligadas a los avances de otros países ; estuvieron al tanto por ejemplo de los avances en los instrumentos italianos promovidos por Bertelli quien llamó " la atención de los físicos sobre los movimientos espontáneos que se observaban en los péndulos tronometros ". Una delegación mexicana asistió en 1901 a la primera exposición de instrumentos sísmicos, lo mismo que una de Rusia, Inglaterra, Alemania, Japón e Italia país que fue el mejor calificado. Se sabía de los primeros experimentos del famoso sismólogo inglés N. Milne quien a finales de la década de los ochenta había pensado de que sería posible registrar los terremotos ocurridos en cualquier punto del globo disponiendo de aparatos de sensibilidad adecuada[5].

Tenemos que desde finales de la década de 1890 hasta 1906, en México se contaba para medir los sismos con una red de estaciones meteorológicas bien comunicadas a partir de la red telegráfica nacional dotadas además en algunos casos de algunos instrumentos " caseros, de manufactura artesanal " o comprados a especialistas y casas especializadas de fuera pero que no daban cuenta de las medidas científicas y homogéneas que empezaban a requerirse en las primeras reuniones de la sociedad sismológica internacional a la que pertenecía México. Esto que en un principio significó una desventaja, en 1906 se convirtió en una oportunidad para la creación de una moderna red nacional a diferencia de lo que sucedió en otros países como Italia y Alemania que tenían ya instrumentos más o menos confiables para medir los sismos pero que quizás

4. M. Pastrana, *Instrucciones para las estaciones meteorológicas del Servicio Metereológico de la República Mexicana*, México, 1904.

5. E. Böse, " Sobre las regiones de temblores en México ", *Memorias de la Sociedad Científica " Antonio Alzate "*, (1902).

debido a esta situación, tardaron más en integrarse a las nuevas redes sísmicas planteadas en las reuniones de las comunidades sismológicas.

PLANEACIÓN DE LA RED

Si queremos hablar de cómo se construyó la primera red sismológica en México tenemos aludir a un doble discurso, el académico y et político que acompaña a la mayoría de los grandes proyectos científicos de nuestra época moderna. Una serie de sucesos marcaron el desarrollo, selección de lugares, instrumentos y personal que constituiría el gran proyecto tan largamente esperado por los sismólogos desde 1880.

Los antecedentes internacionales inmediatos se remontan a los años de 1895-1899 en donde sismólogos ingleses y alemanes (J. Milne entre los primeros y Ehlert y Gerland entre los alemanes) habían hecho diferentes gestiones para lograr organizar del estudio sismológico en todos los países. Para ello hicieron convocaron a los congresos geográficos reunidos en 1895 en Londres y en 1899 en Berlín para estimular la fundación de una asociación internacional sismológica. En estos congresos se logró el nombramiento de una comisión permanente para los estudios de sismología y aprobaron las propuestas hechas por los sismólogos, pero aún así no se logró establecer una asociación internacional. El que trabajó más en este sentido fue el profesor Gerland (profesor de geografía de la Universidad de Estrasburgo) quien consiguió que el gobierno alemán fundara una estación sismológica central pero con las dimensiones y equipamiento suficiente para poder servir de oficina central (construida de 1889-1990) a una organización internacional de estudios sobre temblores, desde aquel momento el profesor hizo todo lo posible para conseguir del gobierno alemán el apoyo para fundar una asociación sismológica internacional. Un primer paso en este sentido se dio en el marco del VII Congreso Geográfico Internacionai celebrado en 1901 del 11 al 13 de abril reuniendo delegados de Austria, Baden, Bélgica, Japón, Rusia, Sajonia, Suiza quienes discutieron las bases de una asociación internacional de estudios sismológicos. En esta primera reunión hubo dificultades para acordar las formas de representatividad de los países y de los científicos pero sobre todo para la organización de la red considerando que Italia ya tenía para entonces una red perfectamente establecida e Inglaterra estaba consolidando la suya, por ello se llegó a la conclusion de elaborar estatutos de una asociación de estados que dejaría a cada uno de ellos en total libertad para organizar su servicio sismológico.

Al año siguiente el gobierno alemán invitó a todos los países considerados como " civilizados " a una Segunda Conferencia Sismológica Internacional en Estrasburgo. México acepto la invitación y mandó como delegado al Sr. José Aguilera, director del Instituto Geológico, a la conferencia que tuvo lugar del 24 al 28 de julio de 1903. Allí se acordó fomentar todas las tareas de la sismología que sólo se pueden resolver por la cooperación de numerosas estaciones

establecidas a lo largo de toda la tierra. La cuota anual de pertenencia a la asociación era de 20.000 marcos (5.000 pesos oro) misma que se destinaba a la
subvención de trabajos relativos a la sismología y para las publicaciones y gastos de la asociación. Prevaleció la idea de la oficina central conectada a la estación central imperial sismológica de Estrasburgo de tal manera que el director
de esta era al mismo tiempo director de la oficina central. Allí se recogerían
los reportes de todas las estaciones en el mundo los cuales se ordenarían y
eventualmente se publicarían.

Los términos de la asociación y su funcionamiento fueron aceptados unáni-
memente por los delegados de Alemania, Austria, Hungría, Suiza, Holanda,
Bélgica, Rusia, Suecia, Gran Bretaña, Portugal, España, Italia, Bulgaria,
Rumania, Japón, estado del Congo, Estados Unidos, México, Chile y Argen-
tina.

Para las mismas fechas en México, algunos científicos como Emilio Böse
(anteriormente lo había hecho don Juan Orozco y Berra) clamaban por el establecimiento de una red científica bien equipada, al respecto Böse escribía :
" En mi concepto todos los estudios sobre temblores en México que se hacen
actualmente son de poca utilidad, porque las observaciones son defectuosas,
especialmente las del tiempo y las del número y dirección de movimientos.
Necesitamos en México una organización metódica, necesitamos una cantidad
de estaciones sismológicas de primero y segundo orden, distribuidas geológi-
camente y tenemos que proveerlas de instrumentas modernos "[6].

El compromiso establecido por el director del Instituto Geológico fructificó,
la primera señal de esto fue que en 1904 como se mencionó, se estableció la
primera estación provisional contigua al observatorio astronómico nacional
dotada de dos péndulos Bosh-Omori de 10 kilos. En 1905 se pudo contar con
otro aparato el " gravímetro " Trifilar de Schmidt. En abril de 1908 fue decretada la creación de un servicio sismológico nacional dependiente del Instituto
Geológico.

El proyecto aprobado para tal fin consistió en el establecimiento de una
estación sismológica central o de primer orden y 52 de segundo.

Aunque para establecer la red sismológica mexicana era importante los
acuerdos científicos de la época, el que el gobierno porfirista financiara tan
costoso proyecto dependía también de la utilidad práctica para la sociedad. Así,
los científicos que apoyaban la red vendían su proyecto argumentando que :
" de la observación continua ya no solamente se obtendrán estadísticas sino la
situación, número y extension de zonas peligrosas, el grado de energía de las
vibraciones que parten de los diferentes focos sísmicos, las direcciones en que
se hace su propagación y el rumbo por el cual hacen su entrada las ondas sísmicas de distinta procedencia en las poblaciones de la república más frecuen-

6. E. Böse, " Sobre las regiones de temblores en México ", *op. cit.*

temente sacudidas por los sismos, para así conocer las direcciones de seguridad, la orientación conveniente para los muros de los edificios, la naturaleza de los materiales de construcción más adecuados ".

La primera propuesta para la red sismológica constaba de una estación central, 5 estaciones de primer orden, estaciones de segundo orden convenientemente distribuidas y observadores corresponsales (al más puro estilo de las estaciones meteorológicas). El conjunto enviarán datos a la estación central. Para la elección de los sitios se tomarían en cuenta los mismos criterios establecidos en Estrasburgo. El objetivo de una red Sismológica era el de adquirir el conocimiento exacto de la sismicidad de un país ; y por tanto, la localización de los epifocos es el primer paso en este conocimiento. Para ello era necesario el establecimiento de al menos tres estaciones sísmicas que permitieran determinar con exactitud este.

LOS APOYOS POLÍTICOS

Pero la trama de la red sismológica tenía que ser tejida con la urdimbre de los apoyos y financiamientos del gobierno porfirista. Gracias a estudios históricos de la época[7], sabemos que el general Porfirio Díaz apoyo de manera importante el desarrollo de la ciencia en México, sobre todo de la ciencia aplicada. Para tener una cabal idea de la conformación de la red, tenemos pues que traer a la escena a estos personajes, porque si bien es cierto que para su establecimiento se retomaron consideraciones académicas, también es cierto que muchos de los criterios para el establecimiento de las estaciones no obedecían, al menos no únicamente, a los dictámenes de los " sabios " de la época.

La Institución que encabezó la propuesta de la creación de la red sismológica fue el Instituto Geológico de México, debido fundamentalmente a los nexos del ingeniero Aguilera con las comunidades geológicas internacionales, a su incorporación en 1904 como Instituto pero sobre todo como país a la Asociación Sismológica Internacional. En este sentido el Ingeniero Manuel Muñoz Lumbier nos relata : " El Instituto Geológico considerando de su incumbencia el estudio de la sismicidad del territorio nacional, presentó al Gobierno un proyecto para el establecimiento de un servicio sismológico, que desde luego fue aprobado, haciéndose las gestiones conducentes para la adquisición del instrumental, construcción de edificios apropiados, selección del personal, etc. Corresponde el mérito de esta obra al entonces director del Instituto Geológico, ingeniero José G. Aguilera, ilustre duranguense de sobra conocido en el mundo por sus trabajos de geología, así como a su colaborador, el ingeniero Juan D. Villarello. El proyecto fue presentado en 1909 "[8].

7. E. de Gortari, *La ciencia en la historia de México*, México, 1980.

8. M. Muñoz Lumbier, " Reseña de sismología ", *Primer Centenario de la Sociedad Mexicana de Geografía y Estadística, 1833-1933*, México, 1933.

Se recogían también clamores de intelectuales de la época como los de Emilio Böse que en 1902 escribía : " Solo en los tiempos modernos es posible la distinción entre los temblores volcánicos y los tectánicos, y esto solamente utilizando instrumentas registradores. En mi concepto todos los estudios sobre temblores en México, que se hacen actualmente son de poca utilidad, porque las observaciones son defectuosas, especialmente las del tiempo y las del número y de la dirección de los movimientos. Hasta ahora no tenemos ninguna organización para observar los fenómenos sismicos, los instrumentas registradores son defectuosos o no funcionan y sobre todo son demasiado pocos. Necesitamos en México una organización metódica, necesitamos una cantidad de estaciones sismológicas de primero y segundo orden, distribuidas geoléógicamente, y tenemos que proveerlas de instrumentos modernos, principalmente de péndulos horizontales con registración mecánica ; necesitamos una verdadera red de observadores y una centralización en los trabajos. Mientras que no podamos conseguir el establecimiento de una estación central de primer orden y varias de segundo orden, no podemos esperar resultados exactos "[9].

Como ya hemos mencionado uno de los proyectos académicos más relevantes que precedió la creación de la red sismológica fue el de la red meteorológica, que siguio patrones académicos similares y que ahora traemos a colación porque nos ayuda a ilustrar el tipo de discurso político que acompañó estos grandes proyectos académicos. Al respecto podemos citar las actas del Primer Congreso Meteorológico Nacional celebrado en 1900 : " La meritisima y modesta Sociedad " Álzate ", a la que me cabe la grandísma honra de pertenecer como su humilde socio honorario, promovio, como todos vosotros sabéis muy bien, esta reunión, tan modesta como ella ; pero en la que figuramos los que, comprendiendo en su verdadero valor la idea que entraña la palabra " patriotismo ", buscamos el enaltecimiento de este bello rincón del mundo que se llama México ; pudiera parecer hasta pretencioso y egoísta el pensamiento expresado ; pero penetrados de su sentido, ampliándolo como debe ser, se verá que es todo lo contrario. México, que durante tres siglos permaneció como esclavo de una de las entonces mas grandes naciones del continente europeo, abrió los ojos a la luz, al grito estruendoso del anciano y venerable cura de Dolores, pareció soltar los apretados lazos que la unían a la madre España, a la entrada a esta capital, de triunfador ejército de las tres garantías, guiado por el libertador Iturbide ; pero ¿ somos realmente independientes desde ese feliz 27 de septiembre de 1821 ? No, señores, quedamos esclavos de la vieja Europa, porque a ella recurrimos siempre por sus artefactos industriales, sus productos generales, dándole en cambio nuestros preciosos metales con que tan abundantemente nos dotó la Providencia : verdaderamente sólo empezamos a salir de esa tutela, a ser verdaderamente libres, a aflojar la cadena de la esclavitud cuando el egregio General Díaz, tomando las riendas del Gobierno nos

9. E. Böse, " Sobre las regiones de temblores en México ", *op. cit.*, 175-176.

ha proporcionado, conduciéndonos tan sabiamente, la anhelada " Paz " de que hoy goza la República ; pues bien. señores, a afianzar esa paz, a entrar en el concierto universal, por el medio científico, a proporcionarnos los medios de bastarnos a nuestras necesidades industriales, agrícolas, higiénicas, tiende esta reunión ; porque extendiendo nuestras redes meteorológicas, uniformando los métodos de observación, reconociendo un centro común, llegaremos a aumentar el material que tan indispensable es para sorprender las leyes que rigen los movimientos de la atmósfera "[10].

En este documento se pedía a los Señores Gobernadores de todos los estados que apoyaran el establecimiento de redes meteorológicas en sus entidades. Estas mismas peticiones se hicieron extensivas cuando se planeo la instalación de la red sismológica de tal suerte que la selección de las dos primeras estaciones Mazatlán y Oaxaca en 1910 y la de Yucatán en 1912 se hizo porque los estados contribuyeron con los gastos y de acuerdo a los criterios internacionales en boga en este sentido podemos citar la selección de los sismógrafos Wiechert sobre los otros que eran conocidos e incluso ya se tenían algunos en México, por ejemplo los Bosch-Omori o los italianos con los que se tenía además una relación muy cercana y eran el modelo a seguir. Se proponía además que los datos de los sismógrafos fuesen comparables y para ello la mejor manera de lograrlo era utilizando los mismos instrumentos en todos los países.

Por último podríamos decir que la continua preocupación del gobierno porfirista por establecer vínculos con el extranjero especialmente con las comunidades europeas ayudó al establecimiento de la primera red sismológica científica en México.

BIBLIOGRAFÍA

E. Addone, " La primera Conferenza Internazionale di Sismología a Strassbourg, 11-13 aprile 1901 ", *Bolletino della Societa Sismología Italiana*, 1901.

R. Aguilar Santillán, " Apuntes relativos a algunos Observatorios é Institutos Meteorológicos de Europa ", *Boletín de la Sociedad de Geografía y Estadística de la República Mexicana*, 1890.

A. Anguiano, " Proyecto aprobado por el Ministerio de Fomento para el establecimiento de un Observatorio Nacional-Astronómico y Meteorológico ", *Anales del Ministerio de Fomento de la República Mexicana*, 1877.

M. Muñoz Lumbier, " La seismología en México hasta 1917 ", *Boletín del Instituto Geológico Nacional*, 36, 1918.

10. *Actas, resoluciones y memorias del Primer Congreso Meteorológico Nacional : iniciado por la Sociedad Científica " Antonio Alzate " y celebrado los días 1, 2 y 3 de noviembre de 1900*, México, 1901, 111-112.

J. Milne, *Seismology*, London, 1908.

A. Offret, " Congreso internacional en México ", *Boletín de la Secretaría de Fomento*, 1907.

Societa Sismología Italiana, " Primo Congresso de Exposizione di Instrumenti Sismici in Brescia (nel settembre 1902) ", *Bolletino della Societa Sismología Italiana*, 1902.

HISTORY OF THE EARTHQUAKE PREDICTION PROBLEM DEVELOPMENT

Svetlana AKHUNDOVA

It is possible to divide the history of the development of ideas and notions of the earthquake prediction into three periods.

1. From antiquity to the middle of the 19th century : direct observation of nature phenomena without their analysis, origin of the idea of the possibility to predict earthquakes.

2. From the middle of the 19th century to the beginning of the 20th century : origin of scientific problems to predict earthquakes in connection with the general standard of knowledge and development of instrumental seismology.

3. From the beginning of the 20th century to the 1980s : searching period of forecasting studies. The 1990s were the period of the analysis accumulated knowledge, the transition to theoretical-physical methodology. There were peculiarities in the correlation between theoretical and experimental works at different stages of the development of the earthquake prediction problem. It has been shown that among the causes of uneven development of this problem there is a chance factor, as catastrophic earthquake. The historical-scientific analysis of the development of earthquake forerunners, the tendencies of their further development, the necessity of the study of social-economical and psychological aspects of this problem in the seismoactive zones are shown.

The prediction of earthquakes consists of two parts : the first is the prediction of its place, the second is its time. This paper deals with the times of the earthquake prediction.

During all its history humanity witnessed some of the most formidable natural phenomenon-earthquakes. Strong underground shocks destroyed the dwellings of people, their property and often took many lives. We traced the history of the development of ideas about the possibility of the earthquake prediction.

Three periods were distinguished from the ancient time till the 1990s.

THE FIRST PERIOD : FROM THE ANCIENTS TO THE MIDDLE OF THE 19th CENTURY

During this period, notions on the possibility of the earthquake prediction arose.

Ancient literature brought us not only information about earthquakes but also observations of people for natural phenomena and linked them with earthquakes. Thus, people observed in the sky a mysterious luminescence, the anxiety of animals, they heard terrible booms before the earthquakes. Ancient astrologists considered that there was a connection between earthquakes and the position of the Moon, the Sun and also with stars, planets and seasons.

Greek philosophers tried to explain earthquakes from the natural and scientific points of view. Thus, Aristotle in the 4th century before Christ stated that earthquakes occurred as a result of the accumulation of air in underground caves resulting in the shaking of the earth's surface.

The first attempt of scientific explanation of the earthquakes in Middle Asia can be found in Abu Ali ibn Sina's papers (he was known in Europe as Avicenna). In his book titled *Kitab Ali-Shifa* or *The Book of Recovery,* dated 980-1037, he dealt with the problems of earthquakes of different magnitude in time and tried to find out the character of influence of the Moon and the Sun on the frequency of earthquakes. He considered that earthquakes depended on the season. Other ancient authors (Khamdullah Kazvini, Muhammed bin Ali-al Khamavi, Rashid-ad-din and others) speaking about the earthquakes which took place in 855, 1042, 1272, 1304, 1640, 1650, 1721 in the Middle East constantly linked them with the weather : that is to say with the eclipse of the Sun, strong wind, hurricane, thunderstorm, lightning, extraordinary snow[1].

Hamdallah Kasvini wrote that astronomer Abu Takhir Shirazi had warned in advance that the earthquake would take place that night in November at ten forty-two. People left the city and exactly that night Tebriz City was destroyed by the earthquake[2].

Even in the remote past, people from their worldly experience learnt that many animals were very sensitive not only to changes of weather but to the approach of natural disasters, for example, earthquakes, too. As a result of centuries-old observations, numerous legends and stories were written. Sometimes they were fantastic and dealt with the animals' behaviour, as if they warned people of a danger. It is interesting that these legends are the same with peoples living not only in different countries but in different continents as well.

In the Middle Ages the perception of the surrounding world was determined by the religious feelings of people. Irrespective of the religion, causes of the earthquake were always linked with the " punishment of God ".

1. As-Suyiti, *Treatise about earthquakes*, Baku, 1983, 78 (translation from Arabic).
2. Hamdallah Kasvini Nuzhat Al-Kulub, *Materials relevant to Azerbaijan*, Baku, 1983, 38 (translation from Arabic).

Egyptian scientist As-sugiti (1445-1505) in his book titled *Treatise About Earthquakes wrote* : " Earthquakes are signs by means of which Allah inspires terror to his slaves... And the Earth is shuddering when someone commits Sin... "[3].

Before the 18th century the observations carried out on earthquakes had a descriptive character. By the end of the century numerous attempts were made to prove scientifically the periodicity of earthquakes and their relation with heavenly bodies and planets.

In the Middle Ages people were gathering information and widening their notions on natural phenomena which were the forerunners of earthquakes. They noticed such features of earthquakes as the change of water level in wells, the emission of gases out of springs and crust fractures, the weak vibration of the Earth before a strong shock, etc.

At the end of the 18th century scientific notions on features which were forerunners of earthquakes seemed to appear. Thus, in 1791, in Calabria (Italy) Italian scientist Pignataro established an interesting dependence between the periodicity of aftershocks and the position of the Moon[4].

The first period, therefore, is characterized by a naïve understanding of the earthquakes' reasons and gathering information on phenomena heralding earthquakes. Also first attempts were made to explain the possibilities of the earthquake prediction from the scientific point of view.

SECOND PERIOD :
MIDDLE OF THE 19th CENTURY - BEGINNING OF THE 20th CENTURY

It is characterized by the rising of the scientific problem of the earthquake prediction in connection with the general level of geological knowledge and the development of instrumental seismology.

Notions on the earthquakes' causes and their prediction are closely connected with the general level of geological knowledge.

At the beginning of the 19th century, ideas on the earthquakes' causes started to spread. In Europe and in Russia, the geology of regions, the causes of mountain building and natural phenomena linked with earthquakes were studied during field trips.

In the middle of the 19th century, French scientist Perrei, on the basis of Pignataro's data and mathematical calculations, determined that earthquakes took place more frequently during New Moon and Full Moon. His successor was Rudolf Falb, who could manage to predict some earthquakes (without showing the place) and caused some sensation around his personality. Later on these

3. As-Suyiti, *Treatise about earthquakes*, op. cit.
4. E. Rothe, *Earthquakes*, Moscow, Leningrad, 1934, 215 (translation from French).

hypotheses was a theory of tides created. Perrei's theory was criticized by many scientists of the world — by French scientist Monteesu de Ballor[5], Russian scientists Orlov[6] and Lagorio[7]. They considered the outside factor (the Moon and the Sun) might be the missing impact if the earthquake in the Earth interior had been already prepared.

In the second half of the 19th century geology was differentiated into separate disciplines. The study of earthquakes began to transform into independent field-seismology. And new epoch of instrumental observations started.

In 1891 in Japan in the province of Mino-Ovari a catastrophic earthquake took place. After that, in 1892, the Japanese government organized an " Earthquakes Studies Committee ". This organization prepared the first governmental Programme aimed at the study of phenomena preceding earthquakes. 1892 became the start point of governmental programmes of the world on the earthquakes' prediction[8]. It is not very well known that after the catastrophic earthquake in Vernyi in 1887, Russian scientist Mushketov[9] took the initiative of setting up a seismic commission at the Russian Geographical Society, which in 1890 prepared a programme of observations for phenomena preceding earthquakes[10].

Thus in the second period in many countries of the world where earthquakes took place, we observe a theoretical-empirical approach to the problem of earthquakes' prediction, that is to say the development of theoretical provisions on the possibility of the earthquakes prediction and methods of research.

THIRD PERIOD (SEARCH PERIOD OF PREDICTING INVESTIGATIONS) :
BEGINNING OF THE 20th CENTURY - 1990s

The beginning of the century was marked by development of seismology. At that time outstanding scientist Golitsyn[11], who lived in Russia, constructed a new device : the seismograph. He also prepared a programme on earthquakes' prediction, caused by the earthquake in Vernyi in 1911. In 1923, a catastrophic earthquake took place in Japan. After that, Japanese scientists paid a great deal of attention to earthquake prediction studies[12].

5. E. Rothe, *Earthquakes, op. cit.*

6. A.P. Orlov, *Earthquakes and their correlation with other phenomena*, Kazan, 1887, 170.

7. A.E. Lagorio, " About earthquakes and their prediction ", *Varshavski Universitet izv.*, 6 (1887), 1-13.

8. E. Rothe, *Earthquakes, op. cit.*

9. I.V. Mushketov, " Earthquakes, their character and ways of observation ", Exploration note to the Questionnaire SPB, *Izv. Russkovo Geographicheskovo obshchestva*, v. 26, 5 (1890), 1-47.

10. S.B. Akhundova, " History of organization of the earthquakes prediction Programmes in Russia and in the USSR ", *Search for geophysical forerunners of the earthquakes in the Caucasus*, Tbilisi, 1987, 3-14.

11. B.B. Golitsyn, " Selections ", *Izd. AN SSSR*, v. 2 (1960), 490.

12. A.V. Nikolayev, " Life and activity of academician B.B. Golitsyn ", *Problems of modern seismology*, Moscow, 1985, 4-9.

In Europe very few scientists were engaged in this problem. American scientists Press and Brace were quite right when they said that not every scientist would take up a risk to express his opinion on the problem of the earthquakes' prediction because they were afraid of their colleagues' blame, for prediction was considered the business of astrologers and fortune-tellers. American scientist Richter called prediction " a wandering little flame " and stated that there was no possibility for the prediction of earthquakes at that time. Research in the field of the relations of earthquakes with cosmic factors went on as well[13].

French scientists More and Bernar, Belgian geophysicist Van Ghils and later on American scientist Knopov and Russian scientist Sytinski tried to reveal the relation between the solar activity and earthquakes[14].

The catastrophic earthquake in Ashkhabad in 1948 stimulated a governmental programme on the prediction of earthquakes[15].

Prominent Russian geophysicist Gamburtsev outlined the programme on the earthquakes' prediction which was the first governmental programme in the world of complex research efforts at the same time theoretical, laboratory and field. At that time (the fifties) in other countries, works in the field of the problem of the earthquakes' prediction had a sporadic character. In Japan they were organized better and were only aimed at the comparison of seismic and some other geophysical data. In the Academy of Sciences of the USSR the problem of earthquakes' prediction was defined as the most important one and this stimulated vast researches in this field[16].

Works of the Soviet scientists in 1949-1955 proved the existence of the earthquakes' forerunners. But owing to the imperfection of the instruments, the lack of data on the physical character of the focus and short period of studies, certain results had not been obtained. Not being scientifically grounded, this problem was excluded from the main themes of the Academy of Sciences. The same sceptic attitude to the problem was observed in other countries.

Agreements made in Geneva in the sixties stimulated the development of seismology on the whole and provided a technical base for further research on earthquake prediction. At the same time radical changes in the system of observations took place, all the instruments recording seismic waves were renewed.

In 1966 a strong earthquake took place in Tashkent. As a result of the retrospective analysis, for the first time in the world, a hydrogeochemical forerunner of the earthquake was determined. There was a sharp increase of radon

13. F. Press, A. Bruce, " Development of the problem of the earthquakes prediction ", *Earthquakes Prediction*, Mir, 1963, 32.

14. A.D. Sytinsky, " About influence of solar activity upon seismicity of the earth ", *Paper of Academy of Sciences of USSR*, 208, 5 (1973), 1070-1081.

15. S.B. Akhundova, *The first state programme on the earthquake prediction (the 40th anniversary of the Ashkhabad earthquake of 1948)*, VINITI Doc., N70-23 1388.

16. A.V. Nikolayev, " G.A. Gamburtsov-Seismologist-Experimentator ", *The development of G.A. Gamburtsov's ideas in geophysics*, Moscow, 1982, 89-102.

content in water before the earthquake. Many scientists linked the fast solution of the prediction problem with that kind of forerunners[17].

The earthquake in Tashkent stimulated the development of prediction works not only in the USSR, but all over the world as well. In 1974 the International Symposium on the Search for the Earthquakes Forerunners that was held in Tashkent, showed that a great progress had been made in the studies of the earthquake prediction and earthquake physics[18]. For the first time in the world an official prediction was carried out. It was made by Chinese scientists the day before the Khaichen earthquake in 1975. The prediction was made according to the complex of the earthquake forerunners[19].

The problem of earthquake prediction was recognized the most urgent problem in geophysics and it had become very prestigious to be engaged in it.

In the seventies scientific papers were characterized by optimism. Many scientists thought that the problem of earthquake prediction would be solved by the eighties. They underestimated the difficulty in solving that problem.

Thus the history of development of the earthquakes' prediction problem has shown that the reason of uneven development of the problem is in the such additional and chance factor like strong earthquakes, that is to say the problem development in different countries after catastrophic earthquakes. And for the last twenty years systematic and planned works have been conducted.

The eighties and the beginning of the nineties were characterized by a system approach to the problem, that is to say all prognostic features of the earthquake were studied at the same time. For that period typical ideas of prediction were taken from field observations. At that time ideas forestalled physical understanding of the studied phenomena. That is why scientists faced the problem of the development of a physical basis for the methods of the earthquakes prediction[20].

With respect to the earthquakes forerunners, one can say that each of them has its own peculiarities of development. Main complex of prognostic features has shown that long ago people knew those forerunners. Many of them, like cosmic, biologic-land electromagnetic[21] suffered qualitative alternations[22]. For the first they were completely denied and later on they were recognized. Many

17. S.B. Akhundova, T.A. Zolotovitskaya, "Geochemical Forerunners", *Hydrogeochemical forerunners of earthquakes*, Moscow, 1985.

18. Tashkent earthquake April 26, 1966, Proceed : G.A. Mavlyanov (chief editor) and others, Tashkent, 1976, 672.

19. D. Davies, "Earthquake prediction in China", *Nature*, 258, 5533 (1975), 286-267.

20. Rikitake, *Earthquake prediction*, Moscow, 3 (translation from English).

21. S.B. Akhundova, R.N. Mamedzade, "The history of the development of ideas about the role of space factors in earthquakes prediction", *Izv. of Azerbaijan Academy of Sciences*, series Nauk o Zemle, 2 (1983), 122-127.

22. S.B. Akhundova, "History of the development of ideas about biological forerunners of earthquakes", *Proceedings of XXIIth-XXIIIth scientific conference*, VINITI Doc., n° 4881-81.

of them such as deformational, seismic and geochemical became leading ones. It should be mentioned that in different stages priority was given to different forerunners.

For the last twenty years much importance has been attached to the social-economic and psychological aspects of the problem of prediction, when the economic calculations of damage from the expecting earthquake and the announced forecast was made, for the prediction can cause material damage as well. The psychological aspect of the problem means possible reaction of people on the announced forecast. In this respect each country should develop a programme of possible behaviour of people after announcing the forecast. And it is necessary from time to time to conduct explanation work in seismic zones. The USA, Japan and China have much experience in finding solutions to psychological aspects of the problem[23].

Thus, in China people trust in prediction and positively react to a false prediction. People in China prefer false alarms to non-predicted earthquakes.

In Azerbaijan false prediction was made twice and it was based on cosmic forerunners. The first prediction was made in 1984 for the fourth of May. It was announced by official authorities that within five days in Baku a strong earthquake would be expected. People were warned about the necessity to take safety measures. Many people left their houses and flats and slept in the yards and even went to the suburbs. The population was not seized by a panic only because people did not trust in the prediction. It has been determined that most people who suffered from earthquakes very well imagined the possibility of an earthquake coming in the area of their residence. And people in Baku did not bear in their minds any strong earthquake. By the end of the week mass media (TV inclusive) repeatedly announced a retreat.

For the second time in the Russian newspapers on the basis of cosmograms, a schedule of expected earthquakes for the summer period, namely June-August 1995, was published. For Azerbaijan it was expected on the twenty-third of August with a 6-7 magnitude. But the prediction was not well founded. People were calm within two months though the latest publications wrote about the possibility of a strong earthquake in Baku. Does not it show that people have got accustomed to a false prediction ?

A new field of seismology called " seismoechology " appeared lately. It deals with the development of prediction of the earthquakes' destructive consequences and their ecological influence.

Thus, we traced the history of the development of the earthquakes' problem starting from ancient times till nowadays.

23. S.B. Akhundova, R.N. Mamedzade, *International seismologic organizations, national and global research on problem of earthquake prediction*, Paper of symposium, Berlin, 1990.

Strong earthquakes in Neftegorsk city in Russia in May 1995 and in Kobe in Japan in January 1995 confused Russian and Japanese seismologists. Lately scientific papers were full of disbelief in a quick solution of the earthquakes' prediction problem.

The thirty years' works aimed at the creation of methods of prediction have not brought the scientists nearer to the solution of the problem. Science cannot exactly predict the time of the earthquake yet.

And now the less and quite agree with American seismologist Mr. Frank Press. He said : " If scientists from all over the world could unite their knowledge and experience, the problem of the exact earthquakes prediction, no doubt, would be solved in the nearest future. Seismologists of different nationalities should unite the results of their work to make a progress on the way to their common goal ".

Tendencies in the Development of Studies of Radioactivity of Natural Environments

Tamara A. Zolotovitskaya

Four stages of research of natural surroundings radioactivity have been distinguished :

1st stage (1920-1950) : gathering information about radioactivity of natural objects, mainly, about waters and mountain rocks. Studies are conducted from time to time.

2nd stage (1950-1960) : tasks of radioactivity research were stipulated both by political and economic conjuncture. They are as follows : search and exploration of nuclear fields, compilation of the catalogue of revealed anomalies, study of radiometry possibilities for the purposes of direct searches of hydrocarbonaceous deposits. This stage is characterized by gathering a great experimental (field) material.

3rd stage (1960-1980) : change of search methodology, analysis of results of field study, followed by theoretical researches to study the genesis of anomalies, tied with deep-seated processes in the Earth's crust, tectonic disturbances, earthquakes, oil deposits.

4th period (1980-1990s) : the Chernobyl catastrophe has caused new scientific direction-radioecology. The aggravation of the environmental situation in consequence of human activity (in Azerbaijan-oil exploration and production) became the subject of radioecology research efforts. In this connection the development of radiometry enters a new qualitative level.

Studies of naturally-occurring radioactivity forces in Azerbaijan reflect all stages of the development of the use of radioactivity.

Being part of the Russian empire and later of the USSR, the territory of Azerbaijan was for many years the site of radiometric research and a location range for the testing of different methods. All this promoted the development of radiometric research at a high scientific-technical level in accordance with tasks which were set both by political and economic situations.

One can distinguish 4 stages of development.

1ˢᵗ STAGE : 1900 TO 1950

The studies of naturally-occurring radioactivity for cognitive purposes were carried out, and the first information on the radioactivity of oil strata waters appeared ; the first laboratory analyses of radioactivity of rocks and minerals were conducted as well.

Radioactivity was measured at that time mainly by the ionisation method with the help of leaflet-like and later thread electroscopes. Despite the fact that the quantitative evaluation of radioactivity was not quite exact and consistent, in that period the foundations of radiometric survey were established, the first attempts were made for the search for oil fields with the help of radiometric methods, first information on the possible relations between geological structure and field of natural radioactivity was obtained.

2ⁿᵈ STAGE : 1950-1960

In the forties there was an improvement in the methods of recording ionizing radiations. Instruments with counters like the Geiger-Muller and compact radiometers appeared. Research became more mobile and an urgent demand for uranium appeared in connection with the production of nuclear weapons for the country.

Uranium which was a waste product of radium production, became the main source of atomic energy. The development of equipment and methods of search for uranium-bearing fields took place in that period.

In Azerbaijan the search for radioactive ores was wide ranging. In the fifties all geological-geophysical field work included the measuring of radioactivity. It was the epoch of the so-called simultaneous or mass search for radioactive raw materials.

One more activity began in Azerbaijan at the beginning of the fifties. A great deal of work was conducted to find out the possibilities of using radiometry for the search for oil-and-gas deposits. As a result it was determined that oil fields were clearly indicated by the anomalous behaviour of the natural gamma-field of the Earth. In Azerbaijan almost all oil-and-gas provinces on-shore and off-shore were studied. In some provinces aerial-gamma ray surveys, including follow-up detailed studies of certain areas by terrestrial methods, were performed. As a result it was determined that all known oil-and-gas structures of the studied regions in Azerbaijan were indicated in the gamma-field by negative anomalies.

Thus, in this stage, radiometric studies in Azerbaijan had an applied character and were aimed mainly at the search for radioactive ores. At that time a catalogue of radioactive anomalies with a brief note of geologic characteristics was prepared. Anomalies around iodine factories and in oil-producing fields were classified as relating to radium source and were of no interest from the point of view of uranic unbearing fields. A large volume of experimental data was gathered based on the results of field surveys and laboratory analyses. However, many problems were not solved relating to the genesis of radio-geochemical anomalies revealed in different regions of the republic.

3rd STAGE : 1960-1980

The further development of radiometry required the development of theoretical questions relating to radiology.

As a result regularities of distribution of radioactive elements in connection with the material composition of sedimentary rocks of the Cenozoic age and the character of spatial variations of gamma-field of the earth's surface in relation to the structural character of the territory and zones of tectonic disturbance were studied. It had been determined that zones of deep faults were reflected in the gamma-field by positive anomalies and that oil occurrences played an important role in the formation of radioactive anomalies.

However, despite such favourable predictions, radiometric methods in the search for oil deposits in Azerbaijan did not develop further methods because at the beginning of the sixties it was recommended that radiometry be used only for the search of tectonic structures. Therefore in 1962 the use of radiometric methods in Azerbaijan with the purpose of searching for oil deposits ceased. And from that time (1965-1966) radiometric research in Azerbaijan acquired a new status. Studies of the pattern of distribution of radioactive fields in connection with geological structure and seismicity began.

In 1965 for the first time in the world an increase of the gamma-field level in the pleistoseist region was for an earthquake of magnitude 7 in Shemakha. That led to systematic observations of variations of the radioactivity in seismogenic zones. Later variability in the radionuclide content in waters and soil gases related to seismicity was determined and on this basis apparatus for earthquake prediction was constructed.

Another important of the radiometric method has been the application of microzonation of land for important government projects, especially in the preparation of building sites.

For the further development of radiometric methods, in 1978, in the Scientific Centre for study named " Geophysics " of the Azerbaijan Academy of Sciences a laboratory " Radioactivity of the Earth's Crust " was established. A new stage in the application of naturally-occurring radioactivity for the solution of different geological problems began. The wide-spread use of gamma-spectrometric devices allowed the study of the distribution of radioactive elements in different structural zones of Azerbaijan and led to a new level in the recognition of the relations between the deep structure of the earth's crust with the processes taking place on its surface. This period is characterized by a change of methods in the solution of different problems. A transition to theoretical solutions in the determination of the genesis of anomalies occurred. It was linked with deep processes in the earth crust, tectonic activity, earthquakes and hydrocarbonaceous deposits.

4th PERIOD : FROM 1986 TO THE PRESENT

On the 26th of April 1986 in the former USSR, in Chernobyl city, a catastrophe occurred, as a result of which there appeared a risk of contamination of vast territories by radionuclides of technogenous origin. In Azerbaijan this caused the beginning and development of a new direction-radioecology.

Thanks to a highly sensitive gamma-spectrometric device which allows the determination of the radionuclide composition of all radioactive materials, the studies of radioactivity of the environment passed to a qualitatively new level and stimulated radioecological research aimed at environmental protection and the limitation of the influence of natural ionizing sources on the community.

In Azerbaijan one more scientific direction-radiostratigraphy is also being developed. Data on the peculiarities of the distribution of radioelements in deposits of the same age within a single sedimentary basin have been obtained. This study is very prospective in correlation of stratigraphic series.

Thus, we traced the development of radiometric studies in Azerbaijan. It should be mentioned that the distinguished periods have been determined not by duration but by the position of science, techniques and state of affairs. Sometimes they accelerate and sometimes they slow down in their development and each new period proceeds from the previous one, demonstrating a progressive development of different directions.

L'UKRAINE ET LA CARTOGRAPHIE OCCIDENTALE
(XIVe-XVIIIe SIÈCLES)[1]

Iaroslav MATVIICHINE

Les mappemondes précoces donnent une idée encore très vague du territoire de l'Ukraine actuelle. Le fleuve *Tanaïs* (Don) y sert de frontière entre l'Europe et l'Asie[2]. Le Don et la mer d'Azov sont mentionnés sur l'une des cartes dans le manuscrit 1119 de la Bibliothèque de Bourgogne à Bruxelles, reproduit schématiquement par J. Lelewel dans son Atlas de 1845.

Les mappemondes du XIVe siècle donnent plus de noms régionaux et les cartes nautiques figurent en détail le contour de la côte des mers Noire et d'Azov. Parmi elles un Atlas nautique de Pietro Vesconte de Genova avec la carte figurant la mer Noire (*Pontus Euxinus*), la mer d'Azov, la Crimée, l'embouchure du Dnipro et du Don, ports. Sa carte maritime de 1327 imita une carte d'al-Idrīsī[3] (1154). La carte du monde connu (1339) d'Angelino Dulcert donne les noms de la *Rutenia*, de la *Ciuita de leo*[4].

---

1. Profitant de l'occasion, l'auteur exprime sa reconnaissance au Prof. Robert Halleux, directeur du Centre d'Histoire des Sciences et des Techniques à Liège dont la recommandation lui valut un grant de son Université pour participer au Congrès de Liège, ainsi qu'au Prof. Alice Payret de Perpignan, fondatrice et présidente de l'association culturelle et artistique *Estem bé*, à Pollestres, pour le soutien dans ses recherches aux Archives de France.

2. *Flum[en] tanais qui diuidit/asiam & europam* dans *Table de Peutinger* (c. IIIe siècle, en copie de 1264) où est également figurée la longue chaîne des Carpates, *Alpes Bastarnice*, les mers Noire et d'Azov et des terres adjointes habitées des *Roxulani Sarmate*.

3. El-Edrisi.

4. L'viv, *cf. Ciuitat de leo* dans l'*Atlante Catalano* (c. 1375) où sont citées aussi d'autres villes en Galicie. *Lelopo* (Leopol) sur la carte de l'Europe centrale (après 1561) de G. Ruscelli, où figurent aussi la *Podolia, Rocatin* (Rohatyn) ; *Sarmatica lago ; Volhinia ; Rossia Rossa, sitomirs* (Žytomyr) sur la *rereteuia f.* (le Teterev), *chiouia* (Kyïv, éloigné du Dnipro), *Nigra silua, Otakou* (Očakiv), delta du *Neper f.* (Dnipro) avec l'île *zagori* ; la partie de bassin du *Nester* (Dnister). Aureli Passaroti, ingénieur italien, a dressé la plus ancienne vue de L'viv, reproduite dans *Civitates orbis terrarum* de Bruin et Hogenberg (1597-1618).

Il connaît Kyïv sous forme de *Chiva*[5]. La carte (*mappamondo*) nautique de Francesco et Domenico Pizigano a presque la même étendue, avec les mers Noire, d'Azov, l'Adriatique et la Méditerranée, la Crimée, le *tanay* (Don), dans toute sa longueur, le *tyrus* (Volga) et plusieurs villes sur leurs bords (12.12.1367)[6]. Les tracés de la Crimée et des côtes des mers Noire et d'Azov sur les cartes maritimes de Vesconte[7] (1311), d'un anonyme (Tunis, 1409), de Bordone (1528), de Crescentius (1596) sont même plus précis que sur certaines cartes russes et turques du XVIIe siècle.

En 1450 environ, Nicolas de Cusa (Krifts, 1401-1464) construit une des premières cartes *modernes*, y compris le territoire d'Ukraine. Sa carte, corrigée et mise à jour, en particulier, grâce à Jan Dlugosz, fut incluse dans les différentes éditions de Ptolémée, et a influencé plusieurs cartographes[8].

Giorgio Giovanni da Venezia (1494) localise *mauro castro* (Bilhorod-Dnistrovs'kyi) sur le Dnister (*flme larllo*) et détaille la partie sud d'Ukraine. Le piémontais Giacomo Gastaldi (1562, 1566)[9] figure les Carpates, la Podolie et place les villes de *chiovia* (Kyïv), *visigrad* (Vyšhorod), *czernigo* (Černihiv). On lui doit des cartes de la Pologne et *Nova Descripcione de la Moscovia* (1562), avec les parties occidentales, nord et orientales de l'Ukraine. Sa mappemonde, publiée à Venise en 1546, fut augmentée et mise à jour par un autre piémontais, Paolo Forlani qui travailla à Vérone et publia à Venise en 1568 *Il vero disegno della seconda parte dil Regno di Polonia*, avec l'Ukraine, enrichie par les noms de plusieurs petites localités, y compris de la région des Carpates (*Tuchla, Sanoch*), des détails de relief et d'hydrographie (toujours avec l'*Amadoca Lago* en Podolie).

Les mers Noire et d'Azov avec la Crimée font le sujet des cartes particulières de Millos (XVIe siècle), Vesconte et Giovanni Maggiolo (1512, 1525)[10],

5. Cf. *Cuiewa* (Thietmar), *Chiue* (Adam de Brême), *chiouia* (Ruscelli, *cf.* n. 4). Plus tard, A. van Westerveld, dessinateur hollandais, a laissé plusieurs vues perspectives et croquis de Kyïv (1643). Une vue de Kyïv du XVIIe siècle est reproduite dans le *Theatrum cosmographico-historicum* [...], publié à Augsburg en 1688 (on y trouve aussi une vue de Caffa). Deux plans des catacombes de Kyïv (au monastère à Laure) de la fin du XVIIIe début du XIXe siècle exécutés par François Rastrelli se trouvent dans les Archives Nationales de France : CP, Sér. N Russie II (*ib.*, du même auteur, une carte exécutée après 1782, des pays compris entre *Oksakow et Cherson*, ainsi qu'un plan d'*Oczacow*, fait avant 1782). Les plans manuscrits plus anciens des mêmes cavernes, datant de 1638, appartient à la Bibliothèque Nationale de Varsovie.

6. B. Palatina à Parma : Ms. 1612.

7. La toponymie dans l'Atlas (1313 ; Bibliothèque Nationale Paris) du piémontais P. Vesconte est plus riche que sur certaines cartes même postérieures et, en particulier, dans l'*Atlante Catalano* ou dans l'atlas de Vencenzo Volcio (1593).

8. En 1562 P. Forlani et Ferando Bertelli, par exemple, ont créé à Venise, d'après Nicolas de Cusa une Carte de l'Europe centrale et orientale.

9. *La discrittione della Transilvania, et parte dell'Ungaria* [...], Bibliothèque Nationale de Madrid.

10. Manuscrit du 7.7.1525, Vesconte et Giovanni Maggiolo : *Carte nautique du bassin de la Méditerranée, des mers Noire et d'Azof (avec la Crimée et les deltas de Dnipro, Don et Danube), de la côte baltique et scandinave*, B. Palatina à Parme, n° 1623.

Agnese (1544), Voltius (1592), Scotti (1593). Une carte dite de Fra Mauro[11], aujourd'hui perdue, servait de base, en 1541 à Candia, pour une carte du crétois Giorgio Sideri dit Callapoda. Sur ces cartes sont figurés le Don, le Dnipro avec une île sur son cours inférieur, et un fleuve à deux embouchures entre le Danube et le Dnipro[12]. Les cartes du XVIᵉ siècle sont suffisamment correctes bien qu'on pût y trouver encore des toponymes de pure fantaisie (*Silva Hircinia, Rissei Montes* chez M. Beneventano[13]). Les italiens Marco Beneventano, Francesco Rosselli et Vesconte Maggiolo[14] ont élargi la présentation des espaces terrestres en comparaison avec l'oecumène de Ptolémée et pour les montrer plus correctement ils avaient choisi les projections en perspective polaire, plus adaptées à la représentation de l'hémisphère septentrional.

La carte de 1548 de Battista Agnese qui, comme Girolamo de Verrazzano (1524), Alfonso de Ulloa à California (1539-1540) et Sebastiano Caboto à Rio de la Plata, se servait de la projection ovale, comprenait pour la première fois, avec plusieurs détails, la partie sud-orientale d'Ukraine, pourtant encore très schématiquement. Dans son *Atlas de cartes maritimes et terrestres*[15], créé vers 1564, Agnese imita l'*Atlas nautique* (1529) du cartographe portugais Diego Ribero au service de l'Espagne à Séville.

Les cartographes portugais du XVᵉ siècle, dont nous ne connaissons que quelques cartes (y compris celle nautique de Jeorge Aguiar de 1492 où les embouchures du Dnipro et Don sont bien représentées), manifestent leur intérêt aux possessions italiennes en Crimée. Une carte maritime anonyme de Lisbonne de 1519 fait flotter dans la mer Noire les bateaux uniquement sous les drapeaux portugais.

Le voyageur anglais Antoni Jenkenson[16] a fait attention plutôt aux parties nord et orientale d'Ukraine dans sa carte : *Russiae, Moscoviae et Tartariae Descriptio*, publiée dans la première édition de l'*Atlas* d'Abraham Ortelius en 1570.

Les autres atlas hollandais, de Gérard et Corneille de Jode (Judeus), ainsi que de Gérard Mercator, donnent une riche information sur l'Ukraine, souvent à travers des cartes des pays voisins, dont certains s'étaient emparés de son territoire.

---

11. La *Carta nautica con elementi corografici prodotta dal laboratorio de Fra Mauro Camaldolese a San Michele di Murano.*

12. Qui semble rejoindre ce dernier (Dnipro) à une distance de la Mer Noire double de celle de l'île mentionnée.

13. Sa carte dans l'édition de Ptolémée, à Rome en 1507-1508, figure l'Ukraine jusqu'au Dnipro. Ptolémée, lui même, désigna l'Ukraine par le nom de Sarmatie.

14. V. Maggiolo, cartographe génois : *Atlas nautique manuscrit*, Naples, 1511 ; *Atlas avec 4 cartes nautiques*, B. Palatina à Parme, ms. n° 1614, 1512 ; *Atlas de 3 cartes nautiques et un planisphère de 1549*, B. municipale de Treviso.

15. Bibliothèque de Médecine à Montpellier, ms. 70.

16. Ainsi que A. Wied (1555, 1594).

Mercator, suivant M. Beneventano et autres, limite la Pologne par ses frontières ethniques, par les fleuves Sian et Buh, et non par le Dnipro. Différents auteurs attribuent le nom de *Russie*, ainsi que plus tard le nom d'*Ukraine*[17] (les deux, écrits sous plusieurs formes) aux différentes parties du territoire d'Ukraine, et ils confondent très rarement la *Moscovie* avec la *Russie* (c'est-à-dire Ukraine) dont les tsars s'approprièrent le nom pour rebaptiser leur pays et prolonger son histoire au fond des siècles. La *Russia*, dans le sens de l'Ukraine occidentale, avec la *Podolia*, est représentée sur la carte dressée[18] en 1559 par le portugais Diego Hommen.

Le nom d'Ukraine était apparu, pour la première fois, sur la carte de 1572, dressée pour le roi de France, Charles IX. Il fut utilisé dans les atlas de Blaeu (1613) et des frères Hondius (1644), et grâce à Beauplan et aux Sansons s'est décidément établi dans la cartographie occidentale. Les cartographes et géographes du XVIe début du XVIIIe siècle utilisent ce nom plus souvent dans le sens du territoire du peuple ukrainien que dans celui de pays de *frontière* (Ukraïna < Okraïna, d'après des linguistes).

Dans son atlas de 1631 J. Blaeu a reproduit une carte de Radziwiłł-Makowski[19] de 1613 d'après les plaques de Gerrits(z), sur 4 feuilles " principales " et 2 feuilles supplémentaires. Sur ces dernières figurent les cours moyen et bas du Dnipro. Avant Beauplan, elles sont un chef-d'oeuvre de la cartographie de cette partie d'Ukraine. Une carte du Dnipro inférieur (de Čercasy jusqu'à l'embouchure), fut publiée à Amsterdam en 1613. Dans la même ville les ingénieurs au service de la Russie, J. Bruce et Corneille Cruys (1657-1727), ont publié, le premier, en 1679, la carte de la Russie européenne, y compris les régions de Kyïv et Braclav, et le second, en 1703, un Atlas du Don, avec 17 cartes, entre autres sur les parties sud et orientale d'Ukraine, mers Noire et d'Azov.

Au milieu du XVIIe siècle Guillaume Le Vasseur de Beauplan, ingénieur militaire français au service du roi de Pologne, en Ukraine, a créé à l'échelon suprême des cartes modernes d'Ukraine. On lui doit la première carte descriptive d'Ukraine, dressée en 1639 (et restée manuscrite), ainsi que les cartes régionales, cartes détaillées du cours du Dnipro, etc. Pour avoir la possibilité de préparer ces dernières, il avait entrepris, en 1639, un voyage le long du Dni-

17. *Cf.* K. Uhryn, *La notion de " Russie " dans la cartographie occidentale (XVIe-XVIIIe siècle)*, Paris-Munich, 1975.

18. Et incluse dans son *Atlas nautique de 1561*, Musée Naval à Madrid, PM 2.

19. Vers 1575 M.K. Radziwiłł, voïvode de Trock et Wilno, commença les travaux préparatifs pour l'exécution d'une grande carte de la Principauté de Lituanie. Pour cela il contacta Maciej Strubicz, des jésuites, des hommes d'Etat en Bélarus et Ukraine. Il semble que la carte ait été prête en 1599, mais son édition fut retardée. L'édition de 1613, la mieux connue aujourd'hui, est due aux graveurs Hessel Gerrits(z) d'Amsterdam, auteur des cartes de la Moscovie, et Tomasz Makowski. La carte contient des textes historiques, ethnographiques et topographiques, dont sur la *Podlesia, ab aliis Polesia [...] volyniae contigua regio*. Sur l'île de la Petite Xortycia du Dnipro est figuré le château de Dmytro Vyšnyvec'kyj, chef des Cosaques.

pro, en faisant les mesures nécessaires. En nouvelle rédaction, ces cartes furent imprimées à Amsterdam, et l'une d'elles (ainsi que la carte générale d'Ukraine) est parue dans les Atlas de Johann Blaeu, en 1662 et 1668. Sa carte générale, *Delineatio Generalis. /Camporum Desertorum /vulgo /Ukraina* [...], avec 6 échelles dont *Milliaria Ocrenica*, fut gravée par Wilhelm Hondius à Danzig en 1648. Elle y fut rééditée en 1654, et en 1660 à Rouen, et est connue en plusieurs autres éditions. Dans la même ville, en 1650, a paru *Delineatio Specialis et accurata Ukrainae cum suis palatinatibus* [...]. C'est le plus grand et important travail de Beauplan et il en existe quelques rares variantes[20].

Les cartes de Beauplan ont encouragé fortement la cartographie européenne, y compris l'activité de la famille des Sanson[21], et la cartographie régionale nationale[22]. En 1665 *le Sr. Sanson d'Abbeville* fait paraître à Paris les 5 cartes de 5 régions administratives *d'Ukraine, Russie Noire, Haute Volhynie, Basse Volhynie, Haute Podolie* et *Basse Podolie*, tirées de *La Grande Carte d'Ukraine, du Sr. le Vasseur de Beauplan.*

Ayant corrigé plusieurs erreurs sur les cartes existantes, Beauplan est l'un de ceux qui ont préparé le fondement pour la réforme de la cartographie effectué par Guillaume Delisle à la fin du XVIIᵉ siècle. La famille des Delisle contribuait fortement à la cartographie de la Russie et de l'Ukraine. En particulier, J.-N. Delisle, étant à la tête d'un observatoire et du " bureau de géographie " à St. Pétersbourg, " il s'efforçait de coordonner les levés de cartes dus aux officiers suédois faits prisonniers à la bataille de Poltava et dont Pierre le Grand avait utilisé les connaissances "[23].

20. Voir [V. Kordt] : В. Кордт, *Материалы по истории русской картографии*, I, Kyïv 1899. Selon K. Buczek, il est " de fait la première carte militaire d'un large espace dans la cartographie d'Europe ". La *Delineatio Specialis* fut rééditée dans l'Atlas des Blaëu sous forme des 4 cartes séparées : Kyïovie, Podolie, Braclavie et Pokuttja. A son tour Nicolas Sanson-père d'Abbeville a aussi réédité ces cartes à Paris en 1659, en y ajoutant encore une cinquième : la Volhynie. Dans la bibliothèque municipale de Gdansk sont conservées 12 cartes des pays riverains de la mer Noire exécutées par Beauplan qui devaient lui servir pour sa *Description d'Ukraine*, ainsi que pour la préparation de sa carte générale de Pologne, imprimée seulement en 1934 par K. Buczek. Beauplan, au cours de son séjour de 17 ans en Ukraine (1630-1647), avait fondé plus de 50 bourgades et fortifié Kodak (*cf.* son *Delineatio Fortality Kudak* [...], 1635).

21. Les cartes de Beauplan servaient comme source principale pour G. Delisle, Jo.-B. Homann, Piter van der Aa, M. Seutter, T.K. Lotter, et plusieurs autres. On doit à la famille des Sanson d'autres cartes figurant différentes parties d'Ukraine. Par exemple, en 1665, G. Sanson dressa une carte hydrographique de la Scythie d'Europe, comprenant le territoire d'Ukraine.

22. Les cartes de Beauplan (ainsi que celles des Sanson) et les levés d'instruments entre 1646 et 1792, exécutés sur le territoire des régiments militaires d'Ukraine, furent exploités dans la carte générale d'Ukraine pour une *Carte en détail de l'Empire Russe*, parue en 1801-1804 sur 100 feuilles, à 1/840.000). De même, dans ce but, on s'est servi aussi des descriptions des régions administratives, comme, par exemple, d'*Opysy Kyïvs'koho namisnyctva 70-80 rokiv XVIII st.* (publiées à Kyïv seulement en 1989) et de la " Description topographique du Gouvernement général (*namisnyctvo*) de Černihiv " (1786) d'Opanas Šafons'kyj, Ukrainien qui était à la tête de la Chambre criminelle du Gouvernement général de Cernihiv.

23. E. Doublet, *Une famille d'astronomes et de géographes*, Bordeaux, [1935], 24. Le tsar lui fit demander de commémorer les batailles en Ukraine, sur le Dnistro et Danube en avril 1737. La carte fut publiée par l'Académie de St. Pétersbourg : *Theatrum Belli Ad Borysthenem, Tyram & Danubium Fluvios gesti A° MDCCXXXVIII*, 50,2 x 69,8.

De temps en temps des cartographes étrangers profitèrent des données des levés et des observations astronomiques exécutées en Ukraine. Daniel de (Bokset), par exemple, dressa plusieurs cartes militaires d'Ukraine, y compris celle du Sič Zaporogue, en se servant largement des résultats des levés, exécutés en Ukraine orientale de 1646 à 1755. Grâce aux levés, effectués sur le territoire des régiments militaires en Ukraine occidentale, en 1725 furent déjà prêtes des cartes des régiments de Xarkiv et de Izjum. Les nouveaux levés achevés (1733), ces cartes avaient été refaites. En même temps était exécuté le levé[24] dans le régiment d'Oxtyrka.

En 1764 fut publiée la carte des régiments d'Ukraine Slobids'ka. Vers 1780 on a achevé le levé de la côte de la mer Noire, jusqu'à l'embouchure du Danube, dont les données furent utilisées par Fedir Čornyj pour publier en 1790 à SPb la carte générale de la Crimée.

Au XVIIIe siècle les levés et travaux cartographiques en Ukraine occidentale[25] furent principalement menés au cadre des programmes polonais et autrichiens par les spécialistes des différentes nations.

L'italien Giovanni-Antonio Ricci-Zanoni publia à Paris en 1772 une carte comprenant la Pologne, la Lituanie, Bélarus et l'Ukraine, en 24 feuilles, sur la base des mesures géodésiques. La partie ukrainienne est complétée par les signes des régiments, des *sotnja* (c'est-à-dire une centaine ; escadron de cosaques).

Plus tôt, en 1700, aussi sur la base du levé géodésique, Bartolomeo Folino avait gravé une carte des mêmes pays, publiée à Varsovie[26]. En 1790, F.I. Maire a achevé la publication du premier Atlas de la Galicie. La carte de Transcarpatie[27], créée en résultat de levé, faisait partie de l'Atlas de Hongrie publié à Vienne pendant 1788-1812.

En 1799, dans la même ville, von Metzburg fait paraître les cartes de la Volhynie occidentale et de Xolmščyna (1/864.000).

A partir du XVIIe siècle la cartographie régionale a connu un considérable

---

24. 1/168.000 et 1/336.000.

25. Dont une partie fut nommée par différents auteurs *Roussie Noire* (Sanson), *Roussie Rouge*, Roussie (Seutter, Vindel). Nicola(s) de Fer évoque *Roussie Rouge ou Roussie Noire*, ansi que *Roussie Rouge ou Roussie Polonaise*. On lui doit un plan de " Kamienieck " (1691 ; Coll. d'Anville, Bibliothèque Nationale, Paris, n° 3153). *Cf. Kamieniec Podolski/Ville Forte des Estats de Pologne, et de la /Haute Podolie [...], Paris, chez le Sr. de Fer, 1705.*

26. Sur 16 ff., 1/235.000.

27. Augustin Hirschvogel cartographia la région d'Ukraine Transcarpatienne (après un levé pas suffisamment approfondi) en 1565. En 1607, A. Passarotti cartographia la région de L'viv. En 1684, Martin Stier publia à Nuremberg sa carte, mise à jour, de la Hongrie : *Vermehrte und Verbesserte/Landkarten des Königreichs Un/garn und deren andern angrentzenden Königreichen/ Fürstenthumen und landschafften [...]* (1/1M., 154 x 102, sur 12 ff. ; *A. Baener sculpsit* ; pas graduée), où la région de Transcarpatie est bien figurée. C'est la seconde édition de sa carte viennoise de 1664, dressée sur la base de celle de Lazius de 1554.

développement. Les régions géographiques et administratives d'Ukraine furent aussi cartographiées par plusieurs spécialistes occidentaux[28]. Les mers Noire et d'Azov, ainsi que la Crimée et le sud d'Ukraine en général, sont le centre d'intérêt des militaires et commerçants[29].

Les Carpates (y compris Bukovine et Transcarpatie), la Galicie[30], la Podolie, la Pokutja, la Volhynie, le *pays des Cosaques*[31], enfin la *Nouvelle*

28. Le choix d'ensemble de la représentation des quelques régions d'Ukraine sur une carte est souvent conditionné par la situation politique ou militaire. Selon *L'Avantcoureur*, n° 6 (1770), 82-83), *le Sr le Rouge, Ingénieur-Géographe du Roi*, a publié un *Recueil de dix nouvelles cartes d'un pied sur 18 pouces contenant l'Empire Turc, la Volhinie, en 3 feuilles ; la Podolie ; la nouvelle Servie ; les Cataractes du Neper ; le Kuban, la Circassie, la Cabardie, la Georgie ; le plan d'Oczakow [...]*. Quant à la carte de *Volhinie* il s'agit certainement de celle publiée partiellement en 3 sections à Paris en 1769 par J.L. Le Rouge sur base de la carte régionale dressée en 1762 par F.F. Czaki. Cette carte a servi aussi pour la représentation de la Volhynie par B. Folin, T.Ph. von Pfau, K. de Perthées, J.A. Rizzi-Zannoni. En 1769, la Firme de B.C. Breitkopf, à Leipzig (Lipsia), a édité la (seconde) carte séparée de Volhynie, *Palatinatus /Volhiniensis/ [...]*, d'un auteur anonyme. Dans le second volume d'*Atlante novissimo* (Venise, 1775-1785) de Antonio Zatto est incorporée la carte des *Li palatinati della Russia Rossa Podolia e Wolhynia tratta dall'Atlante Polacco del Sig.r Rizzi Zanoni* (1781 ; *Presso Antonio Zatta* ; gravée par *G. Zuliani* et commentée par *G. Pitteri*).

29. " Plan Manuscrit de la Mer d'Azof et d'une partie de la Mer Noire 1774 " du capitaine Jan Hendrik van Kinsbergen dans la Bibliothèque nationale de Paris (Coll. d'Anville : Ms. 3099) ; *Carte Topographique / de la Crimée, ou de la / Chersonese Taurique, / avec l'Isle de Taman, et le Golfe de Balaclave / dessinée d'après les nouvelles Observations, et les meilleures Car/tes, et revuë Selon la Géographie de Mr Büsching / par C.L. Thomas / Ingenieur et Géographe à Francfort sur / le Mein, chez qui se vend cette Carte / 1788* (titre aussi en allemand ; en carton : Golfe de Balaclave ; chemins terrestres, indications des profondeurs de mer, forteresses en rouge, bonne topographie grâce aux hachures (Bibliothèque nationale, Paris : Ms. Port. 101, Div. 3, n° 5) ; Carte du gouvernement de Tauride, exécutée à Paris en 1788 (Arch. de la Marine : Rés. 55/68). Parmi des cartographes citons D. Berger (1776, '78), Robert de Vaugondy (1769, vers 1780), J.F. Schmid, F.A. Schrämble (1787), G. Baseggio (1788, '99), Dezauche (1788), le Rouge (1788).

30. A Homann et à ses héritiers on doit plusieurs cartes de Galicie, dont une, *Lubomeriæ / et / Galliciæ / Regni Tabula Geographica [...]* (1775), fut publiée dans l'*Atlas geographicus maior [...]* à Nuremberg. Le cartouche est orné par les drapeaux avec les blasons, y compris de *Rusia Rubra* et *Halicz*. La carte indique, entre autre, les mines expliquées dans la légende. Parmi les toponymes on peut signaler : *Roth Reussie. und Pocutie*. Auparavant cette carte fut publiée dans l'Atlas de C. Lotter, plusieurs fois encore par les héritiers de Homann, entre 1775 et 1813, et par d'autres cartographes : *Carte de la / Pologne Autrichienne / Contenant la Russie Rouge [...]*. *Dressée sur l'Exemplaire / des Heritiers Homann 1775. / à Venise. / Par P. Santini 1776. /Chez M. Remondini // Atlas Universal [...]* par Janvier, Bellin, Robert de Vaugondy, Bonne (Venise, 1778 ; chez P. Santini (*Lieues de Russie Rouge et de Pocutie. Meilen 11/4"*). En 1780 l'éditeur et graveur allemand, Tobias Conrad Lotter, fait paraître la carte politique intitulée : *Carte nouvelle des Royaumes / de Galizie et Lodomerie / avec le District de Bukowine / à Augsburg* dans son *Atlas Novus [...]. Augustae Vindelicorum*.

31. Le 27 octobre 1713 l'agent Baluze (Varsovie) informe la Marine française sur l'établissement éventuel de Cosaques en *Ukraine polonaise* (Archives Nationales Françaises : Mar. B. 7.18.154). Dans les mêmes Archives il y a des cartes et plans liés à l'Ukraine (Mar. 6.JJ.88) ; parmi eux un plan manuscrit du XVIIIᵉ siècle d'une bataille des Tartares, Cosaques et Polonais. En 1748 une carte administrative, en couleur, au 1/3M., *Vkrania quae et / Terra Cosaccorum [...]*, fut publiée à Nuremberg dans l'*Atlas compendiarius [...] recognitus et dispositus J.B. Homann*. Une petite carte manuscrite, exécutée après 1770, du Dnipro aux environ de *[...] où se trouve ancienne Demeure des Saprogores ou Zaporowski*, tirée de la carte de Pologne en ff. publiée à Konigsberg 1770, est conservée dans la Bibliothèque Nationale de Paris (Coll. D'Anville : port. 91, n° 3086). Le *governo cosacco* de *Oczakow* est figuré sur la carte de *La Moldavia e la Valacchia* (1788) incorporée dans l'*Atlante geografico* (1788-1797), édité par Pazzini Carli à Siena.

Servie[32], sont les régions géographiques traditionnelles pour les études cartographiques approfondies, dont la production fut utilisée pour la création de cartes administratives. Le Dnipro[33] et le Don sont cartographiés plus souvent que les autres fleuves.

D'anciennes bibliothèques et archives occidentales conservent un riche matériel sur l'histoire de la cartographie de l'Ukraine, y compris des cartes manuscrites (pour la plupart anonymes et non datées). La Bibliothèque nationale de Paris, par exemple, possède un recueil de cartes et plans, dans lequel une collection d'Anville est la plus considérable. D'Anville, lui même, dressa plus de 200 cartes (et amassa plusieurs ouvrages des autres auteurs), parmi lesquelles[34] : *Carte manuscrite des limites entre la Russie et la Turquie en 1709* ; *Carte manuscrite de la Crimée, de la mer d'Azow et environs* et *Nouvelle Carte Manuscrite de la Crimée et des pays qui sont entre Okzakow [Očakiv] et Azof dressée tant sur les anciennes Cartes que sur les relations de deux Agas Tartares et de quelques voyageurs, communiquées par M. de Maure pour M. D'Anville le 19 x^{bre} 1797*. Dans les Archives nationales de France, l'une des sources importantes est le fonds de la Marine[35] dont une grande partie est toujours gardée dans les Archives de la Marine au château de Vincennes. Parmi les manuscrits conservés dans ces dernières on y trouve une carte de l'Ukraine sud (avec la Crimée dont le contour est fortement déformé) ; deux cartes de la Russie européenne (grandes parties de l'Ukraine, avec Podolie, Volhynie, palatinats *Kievensis*, etc.) ; cartes des opérations de la guerre de 1737, avec un fragment de l'Ukraine sud et la Crimée (Rés. 55/23) ; carte de la Crimée (1740 ; Rés. 55/58) ; plan de la campagne de Crimée 1771 (Ms. 1737) ; plan de *La*

32. Carte manuscrite (1762) de *la Nouvelle Servie* au sud de l'Ukraine dans les Archives de la Marine à Vincennes (Rés. 55/41). Un grand manuscrit en couleurs, des Archives des Armées à Vincennes, est intitulée : *Carte du païs compris entre la Mer Caspienne, la Mer Noire et la Mer d'Azow, contenant l'Ukraine, la Volhinie, la Servie, la Petite Tartarie [...] et toutes les nouvelles conquêtes des Russes sur les Tartares jusqu'aux limites de la perse et de la Turquie. Dressée d'après différentes observations et plusieurs cartes envoyées au Ministre des affaires. Par Mr Calon, Capitaine-Ingénieur Géographe des camps et Armées 1769* (Ms. 4.10. B. 239).

33. Dans les Archives de la Marine à Vincennes est conservée une Carte du Dnipro, exécutée en 1765 d'après des cartes du M. de Peyßonnel (Rés. 55/28), ainsi que sa Carte de la Crimée de 1765 (Rés. 55/56).

34. Bibliothèque nationale de Paris, Coll. D'Anville, port. 91, n° 3087 ; n° 3097 (s.d.) et n° 3098.

35. Par exemple, une carte manuscrite de l'embouchure du Dnipro de 1780 qui est conservée dans le Portefeuille n° 35 de la sous-série 6 JJ (A. Nat.). Ce portefeuille contient également : une carte hydrographique manuscrite du confluent du Buh et du Dnipro, et leur embouchure (1780), faite d'après Saint-Priest ; une petite carte donnant la position de Xerson sur la rive droite du Dnipro ; une carte comprenant une partie du cours du Dnipro et une partie de la Mer Noire (1781) ; un plan de la fortification de la ville d'Očakiv, tel qu'on le trouve dans les mémoires politiques et militaires du général de Manstein (1783) (2.JJ.228) ; une carte hydrographique de la Mer Noire (1785 ; en français ; 6.JJ.35bis) ; un mémoire topographique sur les côtes de la Mer Noire, par Laffitte-Clavé (1784), et autres. Outre plusieurs mémoires on doit à Laffitte-Clavé : un plan d'Očakiv avec projets (1784) et un plan de profils d'Očakiv (1784), un plan du château de *Kodjabej* (1784), un plan du fort de *Hassan Pacha* (1784), etc. (2.JJ.233)

ville de Kamenies (où sont placées : *Porte du costé de la Pologne ; Porte qui regarde la Russie ; Églises qui ont esté changées en Mosquées* ; Rés. 52/90)[36].

Aux Archives diplomatiques du Ministère des Affaires Etrangères de France[37] sont conservés les 3 excellents plans picturaux de Bar, de Medžybiž et de Čyhyryn (le seul muni d'une échelle graphique), ainsi qu'une ébauche d'une partie du Dnipro, envoyés par le marquis de Nointel, ambassadeur de France à Constantinople, à son roi, avec une dépêche du 27.1.1678. Exécutés par un ingénieur, officier français au service du roi de Pologne, qui fortifia ces places (d'après cette dépêche), ces plans sont en couleur : vert (arbres, bosquets, forêts, verdure), bleu (eau), jaune (fortifications), rouge foncé (ponts), rouge clair (partie de constructions civiles et de culte), brun (collines), tout avec les teintes, sous lesquelles on voit les traces des crayons. Pour souligner le relief montagneux de la ville de Čyhyryn, le cartographe devient peintre qui utilise les nuances et la combinaison des couleurs brun et vert. La carte (sans titre, sur une partie de la f. 73) du *Boristenes R.*, à partir de plus au nord de *Chircase* (sur la rive droite) jusqu'à la *Mer Noire*, porte les indications graphiques (lignes qui coupent le fleuve) des *Cataractes* dans le coude du Dnipro, plus bas que *Codatchek*. Sur la rive gauche sont en plus figurées la ville de *Crzemienczuk*, ainsi que la partie basse de 3 rivières, pas nommées. La rive droite porte plus de détails (le *midi* est à gauche). La dépêche mentionne 12

36. Cette forteresse fut souvent cartographiée : en 1687 Giacomo Rossi édita à Rome le *Teatro / della gverra / contro il Tvrco. / Doue sono le Piante, e le Vedute delle principalj / Città, e Fortezze [...]*, y compris Pl. 2. *Kamieniec* (graveur : *Io. Iacobus de Rubeis*, 1684). *Cf.* Nicola de Fer dans une note précédente. Dans le même atlas il y a des plans rares de Mukačevo : 1. *Prospetto d'Eleuatione della / fortezza di MunKatsch*. 2. *Pianta Della Fortezz D[i] Mv[n]katsch (1686)*.

37. Turquie, t. 14, *Correspondance politique*. Voici leurs longs titres explicatifs : 1° *Le vray et exact plan ichnografiques dela ville et chasteau de Cheherin en Ukraine [ukraine], demeure ord.re / du général des Cosaques, nouvellement fortifiée par les polonois, aujourd'huy occuper par les Moscouittes, ou le Duc Ramadonoskj general de l'armée / de Moscovie, et en cette qualité commandant des Moscouittes, Cosaques, Tartares de Calmouk et d'Astrakan a fait entrer son secours en faueur de la / garnison du Chasteau, qui estoit de dix mil Moscouittes, Ce qu'il a executé au mois d'Aou dela presente année 1677 : après la deffaitte des Turcs et des Tartares, qui ont perdu leur bagage, artillerie, et grand nombre de gens (31,7 x 18,7)*. Outre des églises (de rite grec), y sont indiqués des *marais, Estang, rampars*, murailles, chemins, bastions, courtines [*Quatre Courtins fabriquées de poutres de Chesne [...] et si dures que les boules de Canon n'y pouuant faire d'impression retournent, Chose incroyable si je ne l'avois vue, ayant esté assiegé un an dans ce Chasteau, par Dorozensko* [Dorošenko], *des portes, maison de Kmilniskj* [Xmel'nyc'kyj] *general des Cosaques* et celle *de Dorozensko*, lui aussi g[éner]*al des Cosaques* (ff. 72v.-73). 2° *Le vray exact et inographiq ; plan de La ville et château de Meidziboz*, titre suivi tout de suite par une légende où, en particulier, *M. Les Moulins* (une construction haute et forte, avec une roue à eau, à côté de la digue et du pont, où le *Bogek* se connecte à l'*Estang*) ; *N. La digue de l'Estang ; O. Les ponts sur la digue ; Q. Riuiere de Bog ; R. Petite riuiere Bodgek ; S. Le Marais ; V. Bois de haute futaye ; X. Terres labourables ; Z. Chemin de l'ukraine*. 29,7 x 31,14. Le Nord est en haut (ff. 74v-75). 3° *Le vray, exact et ortographique /Plan dela ville et Chasteau / De Bar. / A. La ville de Bar à present ruinée, ou sont restées seulement les ruines des Eglises tant Catoliques que grecques ; B. Le Chasteau de Bar fortifié à la moderne [...] ; C. La ville d'aujourd'hui retranchée ; D. L'Estang long d'une lieue et demie, et large d'un quart ; L. Retranchement dela ville ; M. Les moulins* (à eau ; à côté de :) ; *N. La digue de l'Estang ; N. Les deux pons de la digue*. 30,02 x 31,11. Le Midi est en haut (ff. 76v-77).

Cataractes de ce fleuve, dans le voisinage desquelles demeurent les *Cosaques Cheualiers* (f. 79v).

On peut accorder une importance spéciale aux cartes du XVIIIe siècle dont certaines d'entre elles peuvent être datées plus précisément sur la base des données historiques (fondation des villes, traités, changements des frontières, etc.). A titre d'exemple essayons de dater trois cartes manuscrites en couleurs des Archives de la Marine (Rés. 53/20), non graduées que l'on peut nommer hydrographiques, embrassant le territoire à partir de l'embouchure du Danube, Dnistro et Visla jusqu'à la Volga, avec les mers Noire, Azov et Caspienne. Sévastopol' (fondé en 1784) et Odessa (fondée en 1795) n'y sont pas encore figurés. D'autre part, les Cosaques d'Ukraine (Sič fut détruit en 1775) ne sont déjà plus mentionnés. Alors nous pouvons dater ces 3 cartes *après 1778* (la date de fondation de Xerson, indiqué sur les cartes) et *avant 1784*, soit *vers 1780*. Ces cartes, dressées d'après une source commune diffèrent par certains détails. On y voit la ligne rouge représentant le canal projeté entre le Don et la Volga.

Dès le XVIIIe siècle et même auparavant on trouve une quantité considérable de cartes avec orientation thématique : hydrographique, historique[38], politique, ethnographique, géologique[39], de répartition des langues[40], de frontière, sans dire militaire, maritime. A l'époque, les cartes des routes et des postes (ainsi que d'autres orientations) figurant l'Ukraine portaient l'empreinte des pays limitrophes et spécialement de quatre états envahisseurs, Russie, Pologne, Turquie et Autriche-Hongrie. En été 1787, F.Jo. Maire, déjà mentionné, a commencé son *Atlas des Royaumes de Galicie et de Lodomerie [...] Atlas der*

---

38. En 1765, Jean Baptiste Bourgignon d'Anville exécuta une carte pour le mémoire intitulé l'*Examen critique d'Hérodote sur ce qu'il rapporte de la Scythie (Recueil de l'Académie des Inscriptions et Belles-Lettres*, t. 25, 1765, 573). En 1779, Robert de Vaugondy rééddita la *Romani Imperii / occidentis scilicet et orientis /tabula geographica* [jusqu'au lac de Ladoga, le Don, la mer Caspienne et le Golfe Persique]. Cette carte fut déjà publiée en 1757 dans l'*Atlas Universel* de *Robert et R. de Vaugondy* à Paris, ainsi que dans l'*Atlante* de Santini. En 1790, F.F. Rambach fait paraître une carte des possessions grecques, y compris en Crimée : *Mappa / Colonias / Milesiorum / exhibens / delineata / a / F.E. Rambach* (graveur : Pingeling, Hamburg ; ca. 1/8M.), où sont figurés : sud d'Ukraine, *Pontus Euxinus, Palus Maeotis, Olvia*, fleuves, forts, montagnes figurées par hachures.

39. B. Hacquet traversa la Galicie dans les années 1788, '89, '91 et '93. On lui doit *Observations sur les Monts Carpathes* publiées dans le *Journal de Physique*, t. 40, Paris, 1792, 317-318, où on peut lire qu'il a parcouru en 1791 une partie de la chaîne des montagnes qui sépare la Hongrie de la Pologne, des monts *Crapaths, pauvres en métaux, mais en échange plus riche en sel & eaux minérales [...]*, par exemple, *les eaux de Honasla* à vingt lieues de L'viv. Dans le même volume on y trouve l'information de Martynovyč sur le pétrole de Galicie. Les deux savants ont fait des croquis pendant leurs excursions scientifiques.

40. Lambert ten Kate Hermannsz exécuta une carte intitulée *Volk - en Tael Verspreiding over Europa* (gravée par Jacob Keizer ; éditeurs : Rudolf et Gerard Wetstein) pour un livre : *Aenleiding tot de Kennisse van het verhevene Deel der Nederduitsche Sprake [...]*, publié à Amsterdam en 1723. Sur cette carte l'Ukraine est couverte par *Slavoensche Tak, Kimbrische Tak* et *Theutonische Tak*. La carte : *Europa Poly Glotta* dans *Synopsis universae philologiae* de Gottfried Hensel, publié à Nuremberg en 1741 contient le début du texte du *Pater noster* en plusieurs langues et écritures, y compris en *Polonica, Tartarica et Rvssica*, langues en usage sur le territoire d'Ukraine.

Königreiche Galizien und Lodomerien bestehend in einer general Karte dieser beiden Königreiche, und in 10 besonderen Karten der 19 Kreise nach der neuen Eintheilung nebst dem Distrikt der Bukowina (1788-1790). Certaines cartes furent exécutées par L.Ch. Losy von Losenau, ingénieur du *Zolkiewer Kreises*. Entre 1793-96 *Artaria et Co.* réédita cet Atlas, ainsi que des cartes de F. Maire sur la guerre de Turquie (1787-1791), figurant une partie de l'Ukraine (avec la Bukovyna)[41].

Une contribution des intellectuels provenant de l'Ukraine à l'histoire de la cartographie des autres pays pendant la période considérée, peut être illustrée par l'activité de certaines personnes comme par exemple : le père M. Boym qui vers 1655 exécuta un Atlas de Chine[42] ; Hryhorij Orlyk qui, étant en exil à Salonique, dressa dans son journal (1723) le plan des sources sulfureuses dans les environs de cette ville (et donne leur description) ; Vasyl' Hryhorovyč-Bars'kyj, qui pendant son voyage *aux lieux saints* dressa des cartes et plans d'*Alexandria* (1730), de l'Archipel et des Monastères. Jo. Liesganig, mathématicien et passionné par la géodésie, conseiller du gouverneur à L'viv, fut auteur, entre autres, de plusieurs cartes de la Galicie[43] et de l'Autriche. Balthazar Hacquet (1740-1815), *minéralogiste savant*, professeur d'histoire naturelle à l'Université de L'viv (de 1788), s'intéressa à la géologie des Carpates (en Hongrie, en Pologne et en Ukraine ; *cf.* la note 39). On peut reconnaître un intérêt à la cartographie occidentale chez Kyryl Rozumovs'kyj, hetman des Cosaques et président de l'Académie des Sciences à St. Pétersbourg, ainsi que chez plusieurs autres intellectuels d'Ukraine. Ancien élève d'Euler à Berlin, K. Rozumovs'kyj était en correspondance avec le père Antoine Gaubil. Ce dernier, étant en Chine, a maintenu les relations avec des savants d'Europe avec l'aide du président[44]. Grâce à cette amitié l'Académie des Sciences à St. Péter-

41. En 1769 à Paris, selon *L'Avantcoureur*, 1769, 644), une " Carte générale de la Russie européenne, avec les routes, en 2 ff. " vient d'être publiée par Le Sr Moithey. En 1789, le graveur P.-F. Tardieu exécuta une *Carte des Routes de la Crimée Anciennement Chersonèse Taurique* et les gravures pour le *Voyage en Crimée et à Constantinople, en 1786. Par Miladi Craven*. En 1788, H. Benedicti grava et édita *Poste Karte von der Halbinsel Taurien oder Krim [...]* et une autre carte intitulée : *Confluent et Embouchure du Bog et du Dniéper pour servir de renseignements à la carte des limites des trois Empires ou théatre de la guerre de 1787 et 88. entre la Russie et les Turcs*.

42. Myxajlo Boym (1612, L'viv-1659), jésuite, visita le Viêt-nam, ainsi que la Chine où il a préparé l'Atlas cité (ms. au Vatican) et une *Brevis Sinarum Imperii Descriptio*. Sa *Mapa Imperii Sinarum* fut publiée à Bologne en 1661 par B. Riccioli dans les *Geographiae et hydrographiae reformatae libri duodecim*.

43. Josef Liesganig (Graz, 1719 - L'viv, 1799), jésuite, publia en 1770 à Vienne son mémoire : *Dimensio graduum meridiam Vienensi et Hungarici*. Sur la base des mesures faites en 1772-74 il fait paraître, en 1790, la carte en 33 petites feuilles : *Regna Galiciae et Lodomerie [...] nec non Bukovina* (1/288.000 ; cette carte parut pour la première fois, en 1794, selon Jo. Dörflinger (*Die Österreichische Kartographie im 18. und zu Beginn des 19. Jahrhunderts*, I, Wien, 1984, 95). Les autres de ses cartes furent publiées en 1824, 1847, 1860, 1866, 1868-77.

44. En particulier avec J.-N. Delisle, *ennemi acharné* du président du fait qu'il a ramené à Paris des cartes, dressées par lui en Russie. *Je prends la liberté d'adresser par votre vois à M. Birch, secrétaire de la Soc. royale d'Angleterre, un gros rouleau des cartes avec le plan de Péking* (en 2 copies dont une fut destinée à R.), — écrit Gaubil dans sa lettre du 28 avril 1755 à Rozumovs'kyj. A. Gaubil, *Correspondance de Pékin : 1722-1759*, Genève, 1970.

sbourg s'enrichit en 1755 d'*un rouleau des cartes en caractères chinois*. Rozumovs'kyj visita Voltaire et correspondit avec lui un moment. La carte, intitulée la *Partie de l'Empire de Russie comprise en Europe 1759 et 1760*, pour la première et la seconde édition de l'*Histoire de l'Empire de Russie sous Pierre-le-Grand* de Voltaire, fut dressée par J.B. Bourgignon d'Anville.

La cartographie ukrainienne doit également à plusieurs autres cartographes occidentaux. Les sources cartographiques occidentales enrichissent l'histoire de la cartographie d'Ukraine, peuvent servir pour éclaircir certaines pages de son histoire générale et illustrent les anciennes relations de ce pays avec l'Occident.

Geographical Explorations in Russian America (1741-1867), and Their Influence on the Earth's Sciences' Development

Alexei V. Postnikov

The " discovery " and exploration of America from Asia's side have been processes in which many Nations and Peoples were involved. The importance of different nations in these processes has not been equal and depended on regions taxonomy level.

On the local level the natives played the leading role on the initial stage of explorations, which has begun for them more that 40.000 years ago when they populated the isles and coasts of the Northern Pacific. The local traditional geographical knowledge encircled those territories which provided the means of subsistence for those tribes : hunting grounds, fisheries, plots used for collecting herbs, roots, cereals, and so on. These territories were changing with time due to extinction of life providing resources, and alteration of natural conditions.

Natives' " geographical " knowledge reached up to the regional level as a result of their migrations and transactions with the neighbouring tribes. It must be stressed out that, depending on the available scientific data, it is very difficult to verify how wide the geographical horizon of the ancient peoples has been. It is evident that the " space consciousness " has been conditioned by a mode of subsistence, and reached some sophistication in hunting and nomadic societies. Nations that reached the level of a class society, collected such traditional data from natives of regions being colonized, and created on its basis geographical descriptions and maps included into the general world picture of that time.

In our case Russia was such a state, and Russians explored and mapped territory with use of native geographical information that they gained during their advance to the east. Siberia, Kamchatka, the Aleutian Islands, and the coast of Alaska have been colonized by Russia during a relatively long period of time. Due to the slowness of this advance during the sixteenth through the eighteenth centuries, Russians in some ways became included into the natural and ethnic environment, so to say, and native customs, tools, and crafts became in some measure their own. The natives' knowledge of their land was most important for the exploring and settling of new territories by the Russians. From the first steps of their advance into Siberia, Russians learned to use the geographical information obtained by Siberian tribes, the most crucial of which for orienting in that unfamiliar environment were data on native place-names. Thus native toponymy had been looked for, accepted as their own, and preserved, even if in defaced form by the Russians on their cartographic drawings and maps of the seventeenth and eighteenth centuries. Due to this tendency, one may find old native geographical names on modern Siberian, as well as Alaskan maps.

Native geographical information is reflected for the first time in Russian cartographic relics of the seventeenth century, especially in two drawing (maps) of Siberia dating from 1667 and 1673[1]. Those maps have an importance for the global geography feature : they show an " uninterrupted " water belt off the North East Asia coast between the mouths of the Kolyma and Amour rivers. This knowledge was based not only on the information furnished by Deznev (1648) and other Russian explorers[2], but also on numerous facts learned by Russian *zemleprokhodzy* (trail-blazers) from Chukchi and Eskimo peoples.

The Drawing of Siberia (1667) has come to be called the Godunov Drawing, as its legend proclaims that the work was done " under the scrutiny of the courtier and voevode Pyotr Ivanovich Godunov and his aides ". The map depicts an immense territory to the east of the Volga and Pechora, including the whole Siberia and the Far East. The drawing, like most other pre-Petrine maps, is oriented to the south ; it does not indicate the meridians or parallels. The cartographic depiction looks, at first sight, fairly naive, but the work does in fact convey the main features of the river network. Besides the great rivers like the Ob, Yenisei, Lena, Olenyok, Kolyma and Amour, the work shows many of the tributaries and smaller rivers that flow into the ocean. The drawing is accompanied by a Description, a kind of gazetteer which contains information about sources : the drawing was compiled in Tobolsk in 1667, " using the accounts of all ranks of people, who had been in Siberia's towns and stockades, and knew the terrain and the roads and the lands first-hand... and the travelling natives of Bokhara and the Tartar state servants ". The 1673 *Drawing of Siberia up to the Chinese State...* (dated by F.A. Shibanov at 1669-1670), author unknown, has much in common with the 1667 map, but is a little more detailed. The commentary, or gazetteer to this map, which is actually an updated version of the 1673 gazetteer, survived and contains priceless information, which confirms that a voyage by Semyon I. Dezhnev in 1648 around the Chukotka Peninsula, was repeated by Russian sailors who " from the river Kolyma and around the land... and as far as the Rock [Chukotka Peninsula] did sail, and onwards beyond the Rock, to the river Anadyr and there hunted

1. A.V. Postnikov, " K istorii kartografirovania severnoi chasti Tikhogo okeana i Alaski (do 80-kh gg. XVIII v.) " [On the History of Mapping the North Pacific and Alaska (prior to 1780s). *Voprosy istorii estestvoznaniya i tekhniki*, 3 (1996), 108-125.

2. M.I. Belov, *Semen Dezjnev*, Moscow, 1955 ; R. Fisher, *The Voyage of Semen Dezhnev in 1648: Bering Precursor, with Selected Documents*, London, 1984.

walrus tusk ", and they found it " difficult to circumvent " that Rock. It confirms that these maps are convincing enough evidence that Russians, by the 1670s, knew about straits between Asia and America. This knowledge was substantiated not only by the accounts of Dezhnev, but also by many other tales by Chukchi and the Eskimos who said a Great Land existed across the waters from the Chukotka Peninsula.

The global level of geographical data's processing calls for including this information into the scientific view of the world as a whole. So, Russian seventeenth century maps of Siberia generated a great interest abroad. They were copied and used as sources for maps of Asia by many foreign cartographers of that time. The most influential for the future development of Siberia's depiction on foreign maps was the *Map of European and Asian Russia*, compiled in 1690 by Nucleus Witsen. Witsen, as most foreign geographers and cartographers of that time, was not inclined to fully accept the geographical interpretation of the 1667 and 1673 Russian maps with respect to Northeast Asia, and he put on his map two promontories in the direction of the Arctic and Pacific Oceans, which have suggested the notion of a land connection between Asia and America.

Foreign maps of Siberia have been studied in detail by the famous Russian historian of cartography Leo Bagrow, and, during his fellowship at the Hermon Dunlap Smith Center for the History of Cartography, the present author was fortunate to discover a previously unknown map of this kind. The map *Carte générale de la Sibérie et de la Grande Tartarie...* is in an excellent manuscript atlas in the Ayer Collection of the Newberry Library called the *Carte Marines*. This map is an exceptionally rare and possibly unique copy of a Russian map of the late 1670s or early 1680s for which the author could not find any contemporary foreign map which had any genetic connection to it. The map's rendering of the geography of the region rather faithfully copies the features shown on other Russian maps of that period, although there are some significant distinctions. Most vivid are the map orientation with respect to the north, standard for West-European cartography, and an indefinite representation of the north-eastern extremity of Asia. Here the coastline is interrupted by the map's title where the *Kamchatka River* flows. This leaves open the question of a strait between Asia and America.

Peter I had understood well the Russia's unique opportunities of the compilation of reliable charts of the North Pacific regions. The dream of a northern route from Europe to China and India appealed to Peter the Great, who had been enchanted by all European realities and dreams and tried hard to push Russia into their realm. As with many other Russian reformers of more recent times, Peter I was inclined to teach Russia and Russians to live up to European standards, leaving behind all their previous experiences and traditions. So, in case of geography, which was one of his beloved sciences, he had been reluctant to believe the reports and maps of bearded Siberians and their informants — illiterate natives of Chukotka and Kamchatka. However, Peter the Great knew the Siberian maps and the descriptions of the north-eastern extremity of Asia. There is every reason to believe that he distrusted these materials, and thus wanted to clear up the geographical question as to whether Asia and America were joined or separate.

To fulfil this task the First (1725-1730) and the Second (1732-1742) Kamchatka Expedition were organized, led by Vitus Bering, a Dane in Russian service, and Russian Captain Alexei Chirikov[3]. During these expeditions, and voyage under Mikhail Gvozdev and Ivan Fedorov (1721)[4], Russian sailors confirmed Russia's priority in the discovery of the Strait between Asia and America (later on - Bering Strait), and reached America off North and South Alaska's coasts. During these expeditions the first charts of the North Pacific based on hydrographic surveys have been compiled. Both of Bering's expeditions laid a basis of scientific study and cartography of the Aleutian Islands and Alaska, but, on the other hand, they introduced the usual practice (for scientific overseas voyages) of arbitrary place-naming without any concern for native toponyms. In this respect, these expeditions differed very much from those of their uneducated Siberian forerunners who had been very attentive to native place names. On the other hand, due to Bering and his Navy chiefs, these simple Siberian Cossacks, traders, and sailors, from the time of the second Bering expedition on, were given an opportunity to receive education at special schools of navigation, the first of which was opened in Okhotsk on Bering's order. The curriculum of these schools, besides navigational courses, included geodesy and charts compiling. So, in this way, the traditional methods of old Russian cartography became to be transformed by and unified with scientific European geographical methods, with use of co-ordinates of latitudes and longitudes, permanent scale and cartographic projection.

3. On Bering's expeditions see : A.A. Pokrovskii, *Ekspedizia Beringa* [Bering's Expedition], Moscow, 1941 ; N. Nielsen, " Vitus Berings Stordoad ", *Danmarksposten*, 6 (1942) ; L.S. Berg, *Otkrytie Kamchatki i ekspeditcii Beringa (1725-1742)*, Moscow, Leningrad, 1946 ; D.M. Lebedev, *Geografia v Rossii petrovskogo vremeny* [Geography in Russia during the realm of Peter the Great], Moscow, Leningrad, 1950 ; V.I. Grekov, *Ocherki iz istorii russkikh geograficheskikh issledovanii v 1725-1765 gg* [Outline of the History of Russian Geographical Explorations, 1725-1765], Moscow, 1960 ; V.G. Kushnarev, *V poiskah proliva* [In search of the Straits], Leningrad, 1976 ; R.H. Fisher, *Bering's Voyages : Whither and Why*, Seattle, London, 1977 ; G. Barratt, *Russia in Pacific Waters, 1715 -1825. A Survey of the Origins of Russia's Naval Presence in the North and South Pacific*, Vancouver, London, 1981 ; V.M. Pasetzkii, *Vitus Bering*, Moscow, 1982 ; A.A. Sopozko, *Istoriy plaveniya V. Beringa na bote " Sv. Gavriil " v Sevemyi Ledovityi okean* [The History of V. Bering's voyage on board of " St. Gevriil " to the Polar ocean], Moscow, 1983 ; *Russkie ekspedizii po izucheniu severnoi chasti Tikhogo okeana v pervoi polovine XVIII v.* [Russian expeditions for exploration in North Pacific in the first half of XVIII c.], Zbornik documentai sources], Moscow, 1984 ; N.N. Bolkhovitinov, *Rossia otkryvaet Ameriku, 1732-1799* [Russia discovers America, 1732-1799], Moscow, 1991.

4. A. Polonskii, " Pokhod geodezista Mukhaila Gvozdeva v Beringov proliv 1732 goda " [Geodesist Michal Gvozdev's voyage to the Bering's Straits], *Morskoi zbornik*, vol. 4, 11 (1850) ; A.P. Sokolov, " Pervyi pokhod russkikh k Amerike 1732 goda " [The first travel of Russians to America, 1732], *Zapiski Gidrograficheskogo departamenta...* Chast' 9 (1851) ; V.A. Divin, *K beregam Ameriki. Plavania i issledovania M.S. Gvozdeva, pervootkryvatelia' Severo-Zapadnoi Ameriki* [To the America's coasts: Travels and explorations by M.S. Gvozdev - the discoverer of the North Western America], Moscow, 1956 ; L.A. Goldenberg, *Mikhail Gvozdev*, Moscow, 1982 ; L.A. Goldenberg, *Mejzdu dvumia ekspeditciiami Beringa* [Between two Bering's Expeditions], Magadan, 1984.

Over the next four decades, dozens of private companies were formed to send out vessels and hunting crews. Through these voyages, the various island groups were found — from the near Aleutian Islands up to the Alaska Peninsula, the Shumagins. Although these voyages undertaken by the fur traders were very different from those undertaken by Siberian hunters and explorers along rivers and the Arctic coast, voyages which, in the seventeenth centuries, had brought them to the Pacific Ocean, the fur trades made use of this earlier experience, particularly as regards the collection and use of local geographical information, for their voyages across the open sea. They collected and registered on their charts and in their descriptions many Aleutian toponyms, used the kayak and local methods of orientation and navigation. The nautical charts compiled by the fur traders also preserved many features typical of old Russian drawings, since they were usually drawn up without identifying geographical co-ordinates with help of instruments, had no fixed scale or cartographic projection, etc. With the passage of time, cartographic works produced during such voyages acquired certain features typical of charts compiled in the West European cartographic tradition since the navigators on board these vessels had been educated in schools of navigation. However, the geographical names, the most valuable functional component of the contents of these charts, remained in the original local form.

In the 1770s and 1790s scientific expeditions under Captains Petr Krenitsyn and Michael Levashov as well as Captains Joseph Billings and Gavriil Sarychev were carried out to describe islands visited by *promyshlenniki* and to verify their positions on the map of the Pacific according to geographical co-ordinates[5].

5. P.S. Pallas, " O Rossiiskikh otkrytiyakh mejzdu Aziei i Amerikoi " [On the Russian Discoveries between Asia and America], *Mesiyateslov istoricheskii i geograf cheskii na 1781 g.*, Saint-Petersburg, 1781 ; M. Sauer, *An Account of a Geographical and astronomical Expedition to the Northern parts of Russia, for ascertaining the degrees of latitude and longitude of the Mouth of the river Kovima; of the whole coast of the Tshutski to east cape ; and of the islands in the eastern ocean, stretching to the American coast. Performed, By Command of Her Imperial Majesty Catherine the Second, empress of all Russia, by Commodore Joseph Billings in the Years 1785 and to 1794. The Whole narrated from the original papers by Martin Sauer, secretary to the expedition*, London, 1802 ; V.N. Berkh, *Khronologicheskaia istoria otkrytia Aleutskikh ostrovov ili podvigi rossiiskogo kupechestva* [Aleutian Islands' Discovery's Chronology, or the Heroic Exploits of the Russian Traders], Saint-Petersburg, 1823 ; A.P. Sokolov, " Ekspedizia k Aleutskim ostrovam kapitanov Krenitzina i Levashova 1764-1769 gg. " [Captains Krenitzin and Levashov expedition to Aleutian Islands, 1764-1769], *Zapiski Gidrograficheskogo departamenta...* Chast' 10 (1852), 70-103 ; *Russkie otkrytiya v Tikhom okeane i v Severnoi Amerike v XVIII-XIX vekakh* [Russian discoveris in the Pacific ocean and North America in XVIII-XIX centuries], Moscow, Leningrad, 1944 ; J.R. Masterson, H. Brower, *Bering's successors, 1745-1780. Contributions of Peter Simon Pallas to the History of Russian Exploration toward Alaska...*, Seattle, 1948 ; G.A. Sarychev, *Puteshestvie po severo-vostochnoi chasti Sibiri, Ledivitomu moriu i Vostochnomu okeanu* [Voyage by north-eastern Siberia, Ice Sea and Eastern ocean], Moscow, 1952 ; A.I. Alexeev, *Gavriil Andreevich Sarychev*, Moscow, 1966 ; R.V. Makarova, *Russkie na Tikhom okeane vo vtoroi polovine XVIII v.* [Russians on the Pacific Ocean in the second half of the eighteenth century], Moscow, 1968 ; R.V. Makarova, *Russians on the Pacific 1743-1799*, Tr. and ed. by R.A. Pierce and A.S. Donnelly, Kingston, Ontario, 1975 ; I.V. Glushankov, *Sekretnaiy Ekspeditcia*, Magadan, 1972 ; *Russkaia tikhookeanskaia epopeiy* [Russian Pacific Campaign], Khabarovsk, 1979 ; A.I. Alexeev, *Beregovaia cherta.* [Coast Line], Magadan, 1987 ; D.A. Shirina, *Letopis' ekspedizii Akademii nauk na severo-vostok Azii v dorevoluzionnyi period* [Historical outline of the Academy of Sciences Expeditions to the North-East Asia in before revolution period], Novosibirsk, 1983 ; *Russkie ekspedizii po izucheniu severnoi chasti Tikhogo okeana vo vtoroi polovine XVIII v.* [Russian expeditions for exploration in North Pacific in the second half of XVIII c.], Zbornik dokumentov [Documental sources], Moscow, 1989.

In the preparation for and the conduct of these expeditions the Russian Admiralty was paying much more attention to and was showing much more confidence in the descriptions and maps provided by *promyshlenniki* than before. The most experienced *promyshlenniki* took part in them as official members. To survey the coasts of the Alaska Peninsula, Unimak, and Unalaska Islands, kayaks were used as well as native guides and information.

Gavriil A. Sarychev (1764-1831) was able to combine successfully precise scientific methods of surveying and charting with the use of local information. He used his seven years' experience in surveying in the highly unusual and difficult conditions of the northern part of the Pacific Ocean to compile and publish in 1804 the first Russian manual on hydrographic surveying and cartography. His *Rules pertaining to Marine Geodesy*[6] were republished with supplements throughout the whole of the nineteenth century.

The results of the Russian governmental expeditions together with *promyshlenniki*'s data were in many way crucial for creating the objective picture of land mass distribution in the North Pacific, but they were deposited in the Navy archives and would become known to the general public only many years after the expeditions. The resulting charts showed at least the Aleutian Islands in recognizable outlines to modern eyes. These and some earlier charts are clear evidences of the fact that the knowledge of the Russian Admiralty about the Aleutian Islands, western Alaska, and the vicinity of the Bering Strait to the 1770s was relatively dependable. However, the regions of the American coast between the Bering Strait and the Alaska Peninsula, as well as south of the Alaska Peninsula, excluding parts visited by Bering's expedition, were left unexplored.

Exploring voyages led by famous British Captains James Cook (1778-1779) and George Vancouver (1790-1795)[7], as well as Russian around the World expeditions, beginning from the scientific voyage under Ivan Kruzenshtern and Iurii Lisianskii (1803-1806)[8], have had a great importance for further improvement of Alaska's coasts' mapping.

The polar coasts of Alaska have been surveyed and chartered by Russian

6. G. Sarychev, *Pravila prinadlejzashie k Morskoi geodezii, slujzashie nastavleniem, kak opisycat' moryia, berega, ostrova, zalivy, gavani i reki...* [Rules of the Marine Geodesy for instruction on surveys of seas, coasts, islands, bays, harbors and rivers...], Saint Petersburg, 1804.

7. G. Vancouver, *A Voyage of " Discovery " to the North Pacific Ocean, and round the World, in which the coast of North-West America has been carefully examined and accurately surveyed...*, vol. III, London, 1798.

8. [Yu.F. Lisianskii], *Puteshestvie vokrug sveta v 1803, 4, 5, i 1806 godakh po poveleniu ego imperatorskogo velichestva Alexandra Pervogo, na korable Neve, pod nachal'stvom Flota Kapitan-Leitenanta, nyne kapitana 1-go ranga i kavalera Yuria Lisianskogo* [Ordered by His Imperial Majesty Alexander the First, Round the World Voyage on board " Neva " under Lieutenant-captain Yurii Lisiankii (who is now the Captain of the First Rank) in 1803-4-5 and 1806], parts 1 and 2, Saint Petersburg, 1812.

and British expeditions led by Lieutenant Otto von Kotzebue (1815-1818)[9], Commander Frederick William Beechey, (1825-1827)[10], and Hudson's Bay Company employees Peter Warren Dease and Thomas Simpson, (1837-1838)[11]. The Russian-American Company hunting cruises in the north part of Bering the Sea and Polar Ocean waters adjoining to the Bering's Straits supported and complemented these expeditions.

From the 1820s owards, the Russian-American Company and the Imperial Admiralty elaborated a system of perpetual geographical data collection and mapping in Russian overseas colonies during hunting sails and interior trips of the Company's employees. The hunting and exploring expeditions of Petr Korsakovskii (1818), Ivan I. Vasil'ev (1829), Andrei Glazunov (1833), Ivan Malakhov (1838-1839), and especially Lavrentii Zagoskin (1842-1844) have advanced very much the knowledge about Alaska's interior[12]. An *Atlas of the Northwest Coast of America* compiled by Mikhail D. Teben'kov, and published in Saint Petersburg in 1852 introduced to the scientific world imposing results of Russian geographical explorations in the North Pacific and Alaska. This atlas was for years a standard aid to navigation in colonial waters and helpful

9. O. van Kotzebue, *Puteshestvie v Ujnyi okean i v Beringov proliv dlia otyskania severo-vostochnogo morskogo prokhoda predpriniatoe v 1815, 1816, 1817 i 1818 godakh ijzdeveniem ego siyatel'stva grata Nikolaia Petrovicha Pumiantzeva* [Voyage in the Southern Ocean and Bering Straits in search of North-Eastern Passage organized in 1815, 1816, 1817 and 1818 with Count Nikolai Petrovich Rumiyntzev's sponsorship...], vol. 3, Saint-Peterburg, 1821-1823 ; O. van Kotzebue, *Entdeckung-Reise in die Süd-See und nach der Berings-Stresse zur Erforschung einer nordostichen Durchfahrt. Unternommen in den Jahren 1815, 1816, 1817 und 1818, auf Kosten Sr. Erlaucht des Herrn Reichs-Kanzlers Grafen Rumanzoff auf dem Schiffe Rurich*, 3 vol., Weimar, 1821 ; O. van Kotzebue, *A Voyage of discovery into the South Sea and Beering's Straits, for the purpose of exploring a North-East Passage, undertaken in the years 1815-1818, at the expense of His Highness the Chancellor of the Empire, Count Romanzoff in the ship " Rurick ", under the command of the lieutenant in the Russian Imperial Navy, Otto van Kotzebue*, vol. 1-3, London, 1821, Reprint, Amsterdam, New York, 1967.

10. F.W. Beechey, *Narrative of a Voyage to the Pacific and Beering's Strait to co-operate with the Polar expeditions : performed in His Majesty's ship Blossom, under the command of Captain F.W. Beechey, R.N. F.r.s. &c. in the years 1825,26,27,28.* Published by authority of the lords Commissioners of the Admiralty, vol. 1, 2, London, 1831 ; B.M. Gough (ed.), *To the Pacific and Arctic with Beechey : The Journal of lieutenant George Peard of H.M.S. " Blossom ". 1825-1828,* Cambridge, 1973.

11. T. Simpson, *Narrative of the Discoveries on the North Coast of America effected by the officers of the Hudson's Bay Company during the Years 1836-1839,* London, 1843.

12. L.A. Zagoskin, *Peshekhodnaia opis' chasti russkikh vladenii v Amerike. Proizvedennaia leiytenentom L. Zagoskinym v 1842, 1843 i 1844 godakh. S merkatorskoiu kartoiu, gravirovannoiu na medi...* [A description of the part of the Russian Possessions in America, performed on foot by Lieutenant L. Zagoskin in 1842, 1843 and 1844. With a Mercator Map...], vol. 2, Saint Petersburg, 1847 ; " Ueber die reise und entdeckungen des lieutenant L. Zagoskin im Russischen Amerika ", *Archiv für wissenschaftliche kunde van Russland. Her. van A. Erman,* vol. VI, Berlin, 1848, 499-552 ; vol. VII, 429-512 ; L.A. Zagoskin, " Expedition auf dem festlande van Nord-America... ", *Denk-schriften der russischen geograph. gesellschaft zu St. Petersburg,* vol. 1, Weimar, 1849 ; A. Petermann, *Geographissche mittheilungen,* 1857, Heft IV und V, Juli, 211-312 ; M.B. Chernenko, A.G. Agranat, " Ekspedizia leytenanta L.A. Zagoskina po Russkoi Amerike " [Lieutenant Zagoskin's expedition in Russian America], *Priroda,* 9 (1954), 56-62 ; [L.A. Zagoskin], *Puteshestviya i issledovania leitenanta Lavrentia Zagoskina v Russkoi Amerike v 1842-1844 gg.* [Lieutenant L. Zagoskin's Travels and Explorations in the Russian America, 1842-1844.], Moscow, 1956 ; H.N. Michael, *Lieutenant Zagoskin's travels in Russian America 1842-1844 : The First Ethnographic and Geographic Investigations in the Yukon and Kuskokwim Valleys of Alaska,* Toronto, 1967.

in preparation of charts during the first years of the American period. It is also an outstanding historical source and a monument in the history of Russian cartography, not only due to the quality of its charts, but also because Teben'kov had provided it with *Hydrographic notes accompanying the Atlas of the Northwest coasts of America, the Aleutian Islands and several other places in the North Pacific Ocean*. These notes include precious information on the materials which had been used for compiling the charts, the methods of their application, and the history of the geographical exploration of Russian America.

During all these explorations, besides maps and descriptions, a lot of empirical materials on lithosphere, biosphere, atmosphere and hydrosphere have been obtained. In the course of the second Bering expedition Georg Steller collected in 1741-1742 the first scientific data on nature and natives of Southern Alaska. Being forced to winter together with other survivors of the expedition on Bering Island, he has carried out outstanding observations on flora and fauna, including a scientific description of a sea cow (Steller's Cow) — an animal which fully disappeared at to the end of 18th century due to *promyshlenniki*'s unrestrained hunting[13].

In the 1780s Petersburg's Academician Peter Simon Pallas compiled a scientific outline of the geomorphologic structure of Eastern Asia, the North Pacific, and Alaska, and he was the first to propose a close genetic relationship between Asia's and Northwest America's Mountain Systems and the existence of a land bridge in the past in the regions of Bering Strait and the Aleutian Islands[14]. Some of his reasonings are as follows : " The mountain chains that run along the southern borders of Siberia and extend Northeast between Lake Baikal and the Amour fill the whole of the farthest corner of Asia and are broken off at its eastern coasts. The same mountain range that extends toward the Chukchi Peninsula and throws out branches between the streams entering the Arctic beyond the Lena is quite clearly opposite to the corner of America (which, according to the most reports, lies very close to the Chukchi country) and linked with the American mainland by small intervening islands. The other chief arm of the range, which forms Kamchatka, is part cut off toward the sea on the east side of the peninsula, and appears to show a relationship, through the adjacent Bering Island and Copper Island, to the whole chain of newly discovered islands between Kamchatka and America... Thus the Arctic cordillera seems to have two continuations toward the American mainland : one through the Chukchi Peninsula... the other through the much more southerly chain of

13. G.V. Steller, *Iz Kamchatki v Ameriku* [From Kamchatka to America], Leningrad, 1927 ; G.W. Steller, *Journal of a Voyage with Bering, 1741-1742*, Ed. with an introduction by O.W. Frost, Tr. by Margritt A. Engel and O.W. Frost, Stanford, 1988.

14. A.A. Pokrovskii, *Ekspedizia Beringa* [Bering's Expedition], Moscow, 1941 ; P. Pallas, " Erlauterungen über die im Ostlichen Ocean zwischen Sibirien und America geschehenen Entdeckungen ", I. *Neue Nordische Beiträge*, Bd. 1, Leipzig, 1781, 272-303 ; P.S. Pallas, *Zoographia rosso-asiatica, sistens omnium animalium in extenso Imperio rossico et adjacentibus maribus observatorum recensionem, domicilia, mores et descriptiones, anatomen atque icones plurimorum Auctore Petro Pallas...* (3 vols), (Petropoli, in officine Caes. academiae scientiarum impress. 1811, edit. 1831) ; J.R. Masterson, H. Brower, *Bering's successors, 1745-1780. Contributions of Peter Simon Pallas to the History of Russian Exploration toward Alaska...*, Seattle, 1948.

islands extending from Kamchatka... In both directions there may formerly have been a much easier and closer connection by land between the two continents, which was increasingly broken by the constant currents southward from the Arctic, by earthquakes (which still rage in the chain of islands extending from Kamchatka, which are thickly covered with volcanoes), also by great cataclysms and deluges in the remoter past, which may have violently cut off the solid land as well as the islands (which coasts appear equally torn and broken) "[15].

This guess had been fully confirmed in the nineteenth century, especially in ethnographic works by the Russian Orthodox Missionary Father Ivan Veniaminov (known now as, St. Innokentii, Metropolitan Bishop of Moscow)[16]. These views have laid a basis of modern scientific theories about the Earth's surface and its development in the northern part of the Pacific Ocean.

Published in Saint Petersburg in 1839 Ferdinand Wrangell's and Karl Baer's book *Statistical and Ethnographic Notes on the Russian Possessions on the Northwest Coast of America* was an outstanding work on geography, climatology, native tribes, languages, settlements and other aspects of Russian America which facilitated the creation of Geographical Science[17]. It is a work which has remained of importance to the present day.

From 1839 to 1844, scientific work of great value was performed in Russian America by museum collector Ili'ia Gavrilovich Voznesenski, who was sent out by the Imperial Academy of Sciences. Zoologist I.G. Voznesenskii had studied zoological, botanical, ethnological, and geographical phenomena and shipped to Russia a wealth of collections, which has remained of importance to the present day. His extensive notes, still being deciphered, deal with ethnography, botany, zoology, and other fields. Voznesenkii's collections included

15. J.R. Masterson, H. Brower, *Bering's successors, 1745-1780. Contributions of Peter Simon Pallas to the History of Russian Exploration toward Alaska...*, Seattle, 1948, 24.

16. *Tvorenia Innokentia Mitropolita Moskovskogo kniga tret'ia. Sobrany Ivanom Barsukovym* [Works of Innocent Archbishop of Moscow, the Third Book. Collected by Ivan Barsukov], Moscow, 1888 ; P.D. Garrett, *Saint Innocent, Apostle to America*, Crestwood, NY, 1979 ; P.D. Garrett, " St. Innocent and the Mission on the Nushagak River ", *Orthodox Alaska*, VIII, 2 (1979), 13-21 ; G. Afonksy, *A History of the Orthodox Church in Alaska (1794-1917)*, St. Herman's Theological Seminary, Kodiak, Alaska, 1977 ; L. Black, " Ivan Pan'kov, an Architect of Aleut Literacy ", *Arctic Anthropology*, XIV, 1 (1977), 94-107 ; R.S. Ruthburn, " Indian Education and Acculturation in Russian America ", *Orthodox Alaska*, VIII, 3 and 4 (1979), 79-101 ; V. Rochcan, " The Origins of the Orthodox Church in Alaska, 1820-1839 ", *Orthodox Alaska*, III, 1 (1971), 1-23 and III, 2, 1-15 ; J.W. Van Stone, *Eskimos of the Nushagak River : An Ethnographic History*, Seattle, London, 1967 ; V. Basanoff, " Archives of the Russian Church in Alaska in the Library of Congress ", *Pacific Historical Review*, II, 1 (1933), 72-84 ; J.T. Dorosh," The Alaskan Russian Church Archives ", *Quarterly Journal of the Library of Congress*, XVIII, 4 (1961), 193-203.

17. F.P. Wrangel, " Statistiche und Ethnographische Nachrichten ueber die Russischen Besitzungen an der Nordwestkueste von America ", *Beiträge Zur Kenntniss des Russichen Reiches...*, vol. l, Saint Petersburg, 1838 ; F.P. van Wrangell, " The Inhabitants of the Northwest Coast of America ", J.W. Van Stone (transl. and ed.), *Arctic Anthropology*, 6 (2) (1970) ; F.P. Wrangell, " Russian America. Statistical and Ethnographic Information ", Transl. from German ed. of 1839 by Mary Sadouski, Richard A. Pierce (ed.), *Materials for the Study of Alaska History*, 15 (1980).

more than 400 new species of animals and plants[18].

In 1848-1850, Dr. von Grewing, a geologist from the Russian Academy of Sciences published results of the geological and geomorphologic studies which had been carried out in Russian American colonies[19]. Three maps featuring geology and relief published in his book in 1850[20] were the earliest known thematic maps of natural phenomena for Alaska. One more thematic map was published by the Russian American Company in 1863, when the officials of the Company tried to prove its usefulness in order to prolong its monopoly in Russian America. It is an ethno-linguistic map of the Aleutian Islands and the Northwest coast of America, compiled by Lieutenant-Captain Verman of the Russian American Company.

CONCLUSION

Although the history of geographical exploration of Russian America was an example of many peoples' and countries' efforts in studying their environment, Russia played the leading role in this process. The chairman of the Alaska History Project, Professor Richard Pierce, points out that quantitatively, the achievements of the Russians in gathering information about their colonial holdings outweigh those of any comparable area of North America during that period, or for that matter, of any other newly colonized part of the globe, although a very large region and only a small population were involved. The number of Russians in Alaska was never more than about 800, and at the time of the sale, was only about 500, with about 2.000 Creoles, or people of mixed

18. About Voznesenskii and his collections see : I.V. Glushankov, *Sekretnaiy Ekspeditcia*, Magadan, 197 ; M.G. Stepanova, " I.G. Voznesenskii i etnograficheskoe izuchenie Severo-Zapada Ameriki ", [I.G. Voznesenskii and ethnographic study of North West America], *Izvestiya VGO*, vol. 76, 5 (1944) ; B.A. Lipshitz, " Etnograficheskie materialy po Severo-Zapadnoi Amerike arkhivi I.G. Voznesenskogo ", [Ethnographic materials on the North Western America in I.G. Voznesenskii's archives], *Izvestia* VGO vol. 82, 4, (1950) ; E.E. Blomkvist, " Risunki I.G. Voznesenskogo " (ekspeditcia 1839-1849 gg.) [I.G. Voznesenskii's drawings (Expedition in 1839-1849)], *Zbornik Muzeia Antropologii i Etnografii*, vol. 13 (1951), 22, 23, 83, 184, 186 ; R.G. Liypunova, " Ekspeditzia I.G. Voznesenskogo i ee znachenie dlia etnografii Russkoi Ameriki " [I.G. Voznesenskii's expedition and its importance for the Russian American ethnography], *Zbornik Muzeia Antropologii i Etnografii*, vol. 24 (1967) ; A.I. Alexeev, *Il'iy Gavrilovich Voznesenskii*, Moscow, 1977 ; A.I. Alexeev, *Russkie geograficheskie issledovania na Dal'nem vostoke i v Severnoi Amerike. 19 - nachalo 20 v.* [Russian Geographical exploration in the Far East and North America. 19 - early 20 centuries], Moscow, 1976, 17-19 ; A.I. Alekseev, *The Odyssey of a Russian Scientist: I.G. Voznesenskii in Alaska, California and Siberia 1839-1849*, in R.A. Pierce (ed.), transl. by W.C. Follette, *Alaska History*, 30 (1987) ; K.K. Gil'zen, *Il'iy Gavrilovich Voznesenskii*, Petrograd, 1915.

19. C.J. Grewingk, *Die orographische und geognostische Beschaffenheit der N. W. Küste Amerikas mit den angrenzenden Inseln*, St. Petersburg, 1848 ; C. Grewingk, *Beitrag zur kenntniss der orographischen und geognostischen Beschaffenheit der Nord-West-Küste Amerikas mit den anliegenden inseln. Von Dr. C. Grewingk. (Aus den Verhandlundlungen der Mineralogischen Gessellschaft zu St. Petersburg, für die Jahre 1848-1849, besonders abgedruckt)*, St. Petersburg, 1850 ; K.I. Grevingk, *Orografischeskii i geognosticheskii ocherk severo-zapadnogo berega Ameriki i sosednikh ostrovov,* [Orographical and Geological delineate of North-Western coasts of North America and neighboring isles], Saint-Peterburg, 1850.

20. C. Grewingk, *Beitrag zur kenntniss der orographischen und geognostischen Beschaffenheit..., op. cit.*

blood[21]. The Russian American Company's primary purpose was commercial. How, then, was this record of scientific achievement possible ?

The answer would seem to lie in the keen interest of the Imperial government in the American territories, and in the quasi-governmental nature of the Russian American Company. The Company was required to allocate resources in a way that a purely private endeavour would not have done. The nature of the Russian government — an absolute monarchy — made it possible to order, through its ministries and the company head office, many non-commercial measures. And, as was traditional in the Russian Empire, those who fulfilled the orders in its foreign colonies were among the best educated and soundest officials, naval commanders and churchmen. Such brilliant scientists as Sarychev, von Wrangell, Voznesenskii and Venniaminov, to name but a few, were closely connected with Russian America. Many of their concerns did little or nothing for the commercial success of the colonies, and may even hindered the Company's prosperity, but later generations have benefited from their achievements in non-commercial fields.

21. R.A. Pierce, " Russian Exploration in North America ", *Exploration in Alaska : Captain Cook Commemorative Lectures* (June-November, 1978), Anchorage, 1980, 124.

HISTORIOGRAPHY AND TECHNICS.
NEW RESULTS OF GERMAN HISTORIOGRAPHY

Ute WARDENGA

My presentation was originally supposed to be part of an individual symposium of the commission " History of Geographical Thought " ; it should be entitled " Technology and Geographical Thought ". This was broadly announced world-wide and the " call for papers " appeared in due time. The reaction was meagre and finally only four persons (including myself) were willing to speak on this topic.

Probably the over commitment of the colleagues and the holiday season played a role, but this is perhaps not the entire truth. The real reasons must be more profound and they may have to do with the structure of geography and its research.

During my speech I shall try to focus on these reasons. Because of the time restrictions, I can only discuss German geography and its historiography. It is, in my opinion, somehow different from other national geographies both from its contents as well as from its institutional situation. I believe, nevertheless, that some of my results can be applied on a broader scale, as well.

The history of German geographical historiography can be seen as a continuous differentiation along three different parallel lines. The most ancient line comprises works which define geography in a very broad sense ; they analyze the development of the discipline from antiquity until modern times. The related papers comprise, e.g. Oscar Peschel's *Geschichte der Erdkunde*[1] from 1865 which was edited several times as well as Günther's book with the same title which appeared in 1904[2], and the historical introduction to Hettner's methodology of 1927[3] ; they all originated during a time when geography used

---

1. O. Peschel, *Geschichte der Erdkunde bis auf Alexander von Humboldt und Carl Ritter*, München, 1865.

2. S. Günther, *Geschichte der Erdkunde*, Leipzig-Wien, 1904.

3. A. Hettner, *Die Geographie. Ihre Geschichte, ihr Wesen und ihre Methoden*, Breslau, 1927.

to be defined as a general earth science. Following this opinion, geography was not a single discipline *sui generis* but instead comprised all disciplines producing knowledge about the earth, e.g. geology, meteorology, hydrology, biology, ethnology, cartography and the like. Consequently, historiography was broadly dispersed, including general history, history of the discoveries, history of cartography and natural sciences as well as the history of technics.

Following the broad establishment of the discipline at universities during the last third of the 19th century, a particular scientific community developed and a concentration process took place, resulting in a strictly hermetic national geography with highly differentiated internal structures. As a consequence, a second line of historiography developed whereas the former line regressed.

This newer and more recent approach is characterised by the following : the past is primarily discussed by an aspect which is related to the discipline itself. The development of the emerging special disciplines as well as their methods of research were analyzed, the history of institutions was written or the biographies of famous geographers such as Ritter, Humboldt, Richthofen, Ratzel, Hettner, Penck and others were described. In contrast to the line mentioned first, this type of historiography was supported by a large number of geographers who occasionally made contributions to it.

They were mostly related to a distinct occasion or were written at a defined purpose.

The " papers on occasion " include the progressively increasing publications related to persons, e.g. obituaries, papers in appreciation, historical reviews written to celebrate anniversaries of geographical societies, institutes or periodicals.

The " papers on purpose " include contributions which do not have proper historiographic intentions, because they only use historiography as a tool to illustrate certain facts. This comprises introductions to textbooks or research papers and particularly a large number of reviews concerning the history of the discipline which, by applying historiography, either try to continue traditions or to break with them. Therein, historiography is used as a vehicle to transport own ideas or to justify certain research intentions.

The third line, which developed after world war II tried to overcome the inadequacy of the traditional historiography. This was implemented firstly by applying methods of the historical sciences with a preponderance on documentation and the study of archives materials and its critical interpretation.

Secondly the type of historiography which so far had always been related to the discipline itself was abandoned and geography was characterized as one discipline amongst others against a defined political, social and economic background. Finally, a group of specialists constituted which mainly, if not exclusively, dedicated itself to the research of the development of geographical thought.

One of the first amongst the group of specialists in historiography after world war II was Ernst Plewe. He has published several papers in appreciation as well as obituaries ; their style and beauty of language has never been surpassed until now[4]. From Plewe onwards, a line can be drawn to Hanno Beck[5] and Manfred Büttner[6]. They all have contributed to free geographical historiography from its narrow fixation to geography and have thus constituted the relationship towards the international standards of historiography. A similar development can be observed for the former GDR from the works e.g. of Gerhard Engelmann[7] and Max Linke[8].

The most important impact, however, had the habilitation-thesis of Dietrich Bartels[9] and Gerhard Hard[10], which opened a completely new view on the history of the discipline by discussing long term developments as well as external factors.

During the 1970s, many more papers appeared, which now also introduced political aspects in the development of German geography[11], provided systematic interpretations related to the geographical paradigms[12] or delivered acerbically documented analysis of the methodology of landscape geography[13]. All these papers brought about further research during the 1980s which illustrated the history of geographical institutes[14], the development of single sub disci-

4. E. Plewe, " Geographie in Vergangenheit und Gegenwart ", in E. Meynen, U. Wardenga (eds), *Ausgewählte Beiträge zur Geschichte und Methode des Faches*, Wiesbaden, 1986 (Erdkundliches Wissen, 85).

5. H. Beck, *Alexander von Humboldt*. Vol. I : *Von der Bildungsreise zur Forschungsreise 1769-1804*. Vol. II : *Vom Reisewerk zum " Kosmos " 1804-1859*, Wiesbaden, 1959/1961 ; H. Beck, *Geographie. Europäische Entwicklung in Texten und Erläuterungen*, Freiburg, 1973 ; H. Beck, *Carl Ritter - Genius der Geographie. Zu Leben und Werk*, Berlin, 1979.

6. M. Büttner, *Die Geographia generalis vor Varenius. Geographisches Weltbild und Providentialehre*, Wiesbaden, 1973 (Erdwissenschaftliche Forschung, VII) ; M. Büttner (ed.), *Wandlungen im geographischen Denken von Aristoteles bis Kant. Dargestellt an ausgewählten Beispielen*, Paderborn, 1979 ; M. Büttner (ed.), *Carl Ritter. Zur europäisch-amerikanischen Geographie an der Wende vom 18. zum 19. Jahrhundert*, Paderborn, 1980 ; M. Büttner (ed.), *Zur Entwicklung der Geographie vom Mittelalter bis zu Carl Ritter*, Paderborn, 1982.

7. G. Engelmann, *Die Hochschulgeographie in Preussen 1810-1914*, Wiesbaden, 1983 (Erdkundliches Wissen, 64).

8. H. Lamping, M. Linke (eds), *Australia. Studies of discovery and Exploration*, Frankfurt am Main, 1994 (Frankfurter Wirtschafts- und Sozialgeographische Schriften, 65).

9. D. Bartels, *Zur wissenschaftstheoretischen Grundlegung einer Geographie des Menschen*, Wiesbaden, 1968 (Erdkundliches Wissen, 19).

10. G. Hard, *Die " Landschaft " der Sprache und die " Landschaft " der Geographen*, Bonn, 1970 (Colloquium Geographicum, 11).

11. F.-J. Schulte-Althoff, *Studien zur politischen Wissenschaftsgeschichte der deutschen Geographie im Zeitalter des Imperialismus*, Paderborn, 1971 (Bochumer Geographische Arbeiten, 9).

12. U. Eisel, *Die Entwicklung der Anthropogeographie von einer " Raumwissenschaft " zur Gesellschaftswissenschaft*, Kassel, 1980 (Urbs et regio, 17).

13. H.-D. Schultz, *Die deutschsprachige Geographie von 1800 bis 1970. Ein Beitrag zur Geschichte ihrer Methodologie*. Berlin, 1980 (Abhandlungen des Geographischen Instituts - Anthropogeographie, 29).

14. H. Böhm (ed.), *Beiträge zur Geschichte der Geographie an der Universität Bonn*, Bonn, 1991 (Colloquium Geographicum, 21).

plines[15], the history of geography during certain periods, e.g. the Kaiserreich[16], the Weimar Republic[17] and the Third Reich[18].

Regarding the three types of historiography with respect to the relations between geography and technics (which is my main intention here), a declining line is soon recognizable. Within Oscar Peschel's "History of Geography", the history of measurement technics plays a dominant role[19] and Siegmund Günther[20] stresses the technical aspect of the development of cartography, whereas Hettner, in his historical view already observes the changing styles of geographical descriptions[21]. He reflects the history of the discipline as being a cumulative development with his own construct as a result and peak. Since this construct was aimed at separating geography as an individual and at the same time a uniform science from other sciences as well as from technics, technical progress was of minor importance only. This holds true in a still stronger way for the second type of historiography where the hermetic view towards the discipline itself, its parts and institutions is the main subject. There are only a few papers dealing with the topic of technics but they are restricted to problems of research technics ; e.g. the critical positions of Hans Spethmann[22], Oswald Muris[23] and Hans Schrepfer[24] towards liberalistic and positivistic tendencies within geography.

The opening of historiography since the end of world war II should be favourable for a discussion of the relationship between geography and technics. So it is, but it has, primarily, a negative connotation. Gerhard Hard, e.g.,

---

15. K. Kost, *Die Einflüsse der Geopolitik auf Forschung und Theorie der Politischen Geographie von ihren Anfängen bis 1945,* Bonn, 1988 (Bonner Geographische Abhandlungen, 76).

16. H.-D. Schultz, *Die Geographie als Bildungsfach im Kaiserreich zugleich ein Beitrag in ihrem Kampf um die preußische höhere Schule von 1870-1914 nebst dessen Vorgeschichte und teilweiser Berücksichtigung anderer deutscher Staaten,* Osnabrück, 1989 (Osnabrücker Studien zur Geographie, 10).

17. M. Fahlbusch, " *Wo der deutsche ... ist, ist Deutschland !* " *Die Stiftung für deutsche Volks- und Kulturbodenforschung in Leipzig 1920-1933,* Bochum, 1994.

18. M. Rössler, " *Wissenschaft und Lebensraum* ". *Geographische Ostforschung im Nationalsozialismus,* Hamburg, 1990 (Hamburger Beiträge zur Wissenschaftsgeschichte, 8) ; H. Heske, " *...und morgen die ganze Welt... * " *Erdkundeunterricht im Nationalsozialismus,* Giessen, 1988 ; H.-A. Heinrich, *Politische Affinität zwischen geographischer Forschung und dem Faschismus im Spiegel der Fachzeitschriften. Ein Beitrag zur Geschichte der Geographie in Deutschland von 1920 bis 1945,* Giessen, 1991 (Giessener Geographische Schriften, 70) ; G. Sandner, " Recent advances in the history of German geography. A progress report for the Federal Republic of Germany ", in *Geographische Zeitschrift,* 76 (1988), 120-133.

19. O. Peschel, *Geschichte der Erdkunde bis auf Alexander von Humboldt und Carl Ritter, op. cit.*

20. S. Günther, *Geschichte der Erdkunde, op. cit.*

21. A. Hettner, *Die Geographie. Ihre Geschichte, ihr Wesen und ihre Methoden, op. cit.*

22. H. Spethmann, *Das länderkundliche Schema in der deutschen Geographie. Kämpfe um Fortschritt und Freiheit,* Berlin, 1931.

23. O. Muris, *Erdkunde und nationalpolitische Erziehung,* Breslau, 1934.

24. H. Schrepfer, " Einheit und Aufgabe der Geographie als Wissenschaft ", in J. Petersen, H. Schrepfer, *Die Geographie vor neuen Aufgaben,* Frankfurt am Main, 1934, 61-86.

in his habilitation-thesis already mentioned (1970) has essentially analyzed the term " landscape ". By applying linguistic methods, he was able to show that common language connotations of the term have strongly influenced the scientific analyses of the so-called " landscape geography ". In other words : Hard demonstrated the intimate relations of a scientific approach and colloquial language and concluded that " landscape geography " was a kind of proto- or pseudo-science. His results and conclusions were shocking, because landscape geography had been continuously developed towards a general approach of German geographers. It comprised not only the traditional regional descriptions of the so-called *Landes- and Länderkunde* (regional geography), but dominated also large parts of the research in physical and human geography.

During the seventies the results of Hard's research initiated a broad, very emotional and conflict-ridden discussion on the importance of the landscape concept. Within this discussion historical research was a main tool to support the critical view especially of the young generation.

Due to the time restrictions, I can only comment on the main results.

Hans-Dietrich Schultz e.g. has shown that the landscape concept originated from school geography and was thus combined with an institution that university geographers regarded to be not scientific in its tradition[25].

Ulrich Eisel went still one step further. His systematic interpretations regarding structures and contents of the geographical paradigms revealed that the landscape approach structurally developed as a project directed against modernity and hence against industry and technics. In spite of all modernisation trends, Eisel showed that landscape geography always placed man as a concrete-sensuous, regionally-bound individual who either lived in harmony or conflict with the concrete ecological nature[26]. Furthermore it became clear that the position against technics which was held by the classical German geography was not accidental but originated from the landscape approach and its affinity to non-industrialized, pre-modern lifestyles. Klaus Kost e.g. has shown by his dissertation entitled " The influences of geopolitics on research and theory of political geography " that the break-through of geopolitics and political geography during the 1920s was nothing else but a counter reaction against progress, technics and industrialisation[27]. Horst Alfred Heinrich obtained similar results from a quantitative analysis of periodicals ; by this, he was able to demonstrate the affinity of geographical research and fascistic ideas[28].

25. H.-D. Schultz, *Die deutschsprachige Geographie von 1800 bis 1970...*, op. cit.

26. U. Eisel, *Die Entwicklung der Anthropogeographie von einer " Raumwissenschaft " zur Gesellschaftswissenschaft*, op. cit.

27. K. Kost, K. Kost, *Die Einflüsse der Geopolitik auf Forschung und Theorie der Politischen Geographie von ihren Anfängen bis 1945*, op. cit.

28. H.-A. Heinrich, *Politische Affinität zwischen geographischer Forschung und dem Faschismus im Spiegel der Fachzeitschriften...*, op. cit.

After a period of rejection during the eighties the results of these historical research projects are meanwhile broadly accepted, even by some of the elder university geographers who were sceptic at the beginning. According to the historical view constructed since the seventies at the moment, a vast majority of German geographers believes that the landscape approach was the one and only paradigm of the classical German geography which came to an end during the seventies. Secondly especially those 40, 50 and 60 year old geographers, who have studied at an institute of the former Federal Republic of Germany think that the landscape concept produced conservative and premodern views of the world. Recently Benno Werlen for example has used this interpretation to develop his social geographic approach which is — in contrast to the classical German geography — oriented at the problems of modernity and modernization[29].

During the last five years, however, some historians, especially Hans-Dietrich Schultz and I myself began doubt these general opinions. Apart from each other we started from a careful research on the geography of the Kaiserreich, the period from 1871 up to the end of world war I. In several works on Hettner's geographical concept and the situation of German geography in the first three decades of our century, I was able to show that landscape geography was not working as a paradigm since the second half of the 19th century as generally supposed, but succeeded only after the traumatic events of world war I[30].

It also became apparent that geography during the entire 19th century was much more progressive and directed towards technics than those historians believed who had step by step reworked the structure of the landscape concept. Within a Festschrift for Gerhard Hard still to appear, Hans-Dietrich Schultz demonstrates in an impressive manner, how the classical German geography started at the beginning of the 19th century its career from a position at the side of technical and industrial progress and then, stepwise during the 20th century, especially after world war II, developed towards an attitude that was directed against technics, against industry, against progress and against modernization[31]. The conclusion to be drawn from our results is, that the negative atti-

29. B. Werlen, *Sozialgeographie alltäglicher Regionalisierungen*. Vol. 1 : *Zur Ontologie von Gesellschaft und Raum*, Stuttgart, 1995 (Erdkundliches Wissen, 116). Vol. 2 : *Globalisierung, Region und Regionalisierung*, Stuttgart, 1997 (Erdkundliches Wissen, 119).

30. U. Wardenga, *Geographie als Chorologie. Zur Genese und Struktur von Alfred Hettners Konstrukt der Geographie*, Stuttgart, 1995 (Erdkundliches Wissen, 100) ; U. Wardenga, " Nun ist Alles, Alles anders ! " Erster Weltkrieg und Hochschulgeographie, in U. Wardenga, I. Hönsch (eds), *Kontinuität und Diskontinuität der deutschen Geographie in Umbruchphasen. Studien zur Geschichte der Geographie*, Münster, 1995 (Münstersche Geographische Arbeiten, 39) ; U. Wardenga, " Geschichtsschreibung in der Geographie ", *Geographische Rundschau*, 47 (1995), 523-525 ; U. Wardenga, " Geographie als Chorologie - Alfred Hettners Versuch einer Standortbestimmung ", in D. Barsch, W. Fricke and P. Meusburger (eds), *100 Jahre Geographie an der Ruprecht-Karls-Universität Heidelberg (1895-1995)*, Heidelberg, 1996, 1-17.

31. H.-D. Schultz, " Von der Apotheose des Fortschritts zur Zivilisationskritik. Das Mensch-Natur-Problem in der klassischen Geographie ", *Festschrift für Gerhard Hard*, Kassel, 1997.

tude towards technics is obviously only a short and recent interlude in the performance of German geography. Regarding historiography, however, our results show that one has to be extremely careful in the interpretation of historical texts. This sceptical finale should, however, be taken with caution. Historiographers always remain bound to their own time and they will always direct questions of their times towards history. But I think in future geographical historiography must develop independent categories to become an authority able to reflect its observations beyond an only intrinsic disciplinary view. As long as historiography is used as a tool to legitimate present research interests and as long as the resulting pictures are trimmed towards the most recent actuality, we will construct historical views in which the independent, the different and the original of the past disappears in favour of an only presumed tradition.

THE TRANSFORMATION OF THE IDEA OF SPATIALITY IN THE 20th CENTURY GEOGRAPHY

Grigoriy KOSTINSKIY

The aim of this paper is to analyse the transformations of the idea of spatiality throughout 20th century geography. The idea of spatiality is chosen since it has an outstanding significance for this discipline. The extreme importance of this idea particularly for geography is caused by the fact that, on the one hand, it holistically grasps the object of geographical study in the form of spatial singularity (" geospace ") and, on the other, determines the basis of its research method (" spatial approach "). The idea of space is perceived within this discipline as a " mother concept " of human geography that fulfils a significant explanatory function and manifests specificity of geographical research. It is regarded as one of the generating structures of geography, that is such schemes of thought-and-action, that delimit (construct, design) its subject. The profound theoretical discussion of space as a basic category of geographical knowledge began in the 1970s with the books of Yi-Fu Tuan and R. Sack[1].

The classic 19th century geographical research program found its most distinct and intelligible form in the so-called chorological (based on the idea of space) conception by Alfred Hettner. Since the very spirit of geography was to encompass the whole set of the processes concentrated in one locality, this program urged to consider things and processes on the earth surface as filling of space (space as a " container " or " reservoir "). On the eve of the 20th century geography rested upon the vision of a single and homogeneous space, which was dominating in culture and science. Nevertheless, soon this dominating idea was subjected to substantial revision by physics but most profoundly by philosophy (Husserl, Heidegger).

In the late 1950s parallel with idea of space as a container the idea of space as a structure appeared. It stems from the conception of isomorphism of differ-

1. R. Sack, *Conceptions of space in social thought,* Minneapolis, 1980 ; Y.-F. Tuan, *Space and place. The perspective of experience,* London, 1977.

ent spheres of reality brought with the system analysis. The third mode of spatial understanding views space as an image. This mode of understanding was born in the 1970s together with a humanistic stream in science, emphasising that a human being is both a maker and a keeper of images.

THE SIGNIFICANCE OF SPACE AND SPATIAL NOTIONS FOR GEOGRAPHY

It is very symptomatic that to the notion of space the definitions of the most important categories of geography (place, territory, region, and some others) are attributed[2].

In the flow of speech geographers (and not only they, of course) mix the concepts of space, place, territory, region, area, landscape and some others, which often act as synonyms. Such a " confusion ", to be sure, is not a result of ignorance but has a very serious ground. These universal notions under scrutiny are so deeply connected and conjugated, that they are able to replace semantically each other in our consciousness. This certificate their generality with a certain " historical ancestor ", and a generic unity.

A generic closeness, propinquity and distinct interchangeability of spatial concepts, however, does not free us from the necessity to reveal their specific character, specificity in the logical and historical formation of each of them, and from an attempt to explain on what concrete account they are capable to replace each other so easily.

My concern in this article is to offer insight to the relations between basic geographical notions — space, place, territory and region in the framework of a single conception of spatiality. This set of four notions mentioned is very close to that offered by E. Relph, but with one exception : he includes in it a landscape instead of territory[3].

Traditionally, since Kant and Lessing there is a conventional division of sciences and arts into spatial and temporal and this distinction has been very important for modernity.

Geography was defined by Kant as a spatial science. Space as the idea and concept has come and fixed in geography in a double way : 1) as the integrity of the continuum of geographical objects (A. Hettner), and 2) as the property of extension of objects (especially since the time of formation of spatial analysis school in the late 1950s). At the turn of the 19th century the conception of geography as the science studying organisation of space and the similar conception of architecture as the art of space formation have arisen. Note, that the

2. Compare their definitions, for instance, in : R.I. Johnston, D. Gregory and D. Smith (eds), *Dictionary of Human Geography*, 3rd edition, Oxford, 1994. At the absence of a better term we call these notions " spatial concepts ".

3. E. Relph, " Geographical experiences and Being-in-the-World : the phenomenological origins of geography ", in D. Seamon, R. Mugeraurer (eds), *Dwelling, place and environment*, Dordrecht, Boston, 1985.

interest to space amplified in the periods when the question about the subject of geography and its specific character was becoming acute.

The notion of space in geography could not but become an extremely important explanation model, as far as, on the one hand, it has successfully expressed the unity of the whole, and on the other, enabled to introduce operational (geometrical) explanation into geography.

Though there are opponents in consideration of space in isolation from time, who call to examine social processes as the unity of space and time, this does not interfere them in practical investigations to think of space and time separately from each other.

The idea of spatiality functions differently under positivistic (naturalistic) and phenomenological (culturological) attitudes of consciousness[4]. It is necessary to note that the concept of space in geography has developed under the dominance of naturalistic attitude of geographical consciousness. And nowadays, in geography this mode of understanding of space is still dominating. Geographers continue to think of it mainly as of a physical space (albeit in two variants — as absolute and as relative).

The most widespread and stable understanding of space is that of a container (a reservoir). Space is seen as an emptiness, capable to serve as a place of presence for things or processes. In this case the thought concentrates on such parameters as length, width, area, volume. Then, the Earth space (geospace) is thought as the only one (since the Earth is unique) and it receives the status of an objective (and/or real, physical) one.

Another version of the naturalistic approach to space (signifying, however, a withdrawal from its orthodox variant) — a recognition alongside with a real (physical) space also a number of reflexive spaces — usually a perceptive and a conceptual one. Such an approach is forced to construct a sort of bridge between a pure " material " (corporeal) space and a pure " ideal " space. This bridge looks as a series of intermediate spaces — otherwise the metaphysical dualism cannot be overcome (it is possible to call this situation " a metaphysical trap ").

If the adherents of such an approach qualify the space, in which various processes on the Earth, outside human consciousness, occur, as real, then a perceptive space is a modification of real space, reflected by human perception (senses), and conceptual — a deeply reflective speculative space of abstract models and concepts. However, such a distinction between spaces is poor as having obvious defects, since even the real space is an abstraction, for it is given to us either in a perceptive, or in a conceptual form. It is also not clear why a perceptive space is independent from a conceptual one, though it is

4. J. Pickles, *Phenomenology, science, and geography*, Cambridge, 1985 ; G.D. Kostinsky, " The attitudes of consciousness and different traditions in geography ", *Izvestiya Akademii Nauk. Seria Geogr.*, 5 (1990), 123-128 (in Russian).

inevitably corrected by thinking. Perceptual and conceptual spaces have been renounced a high-grade " true " status.

H. Couclelis and N. Gale distinguished quite a big number of spaces of different kind — a pure Euclidean, physical, sensomotor, perceptual, cognitive, and a symbolical one[5]. In her newer work H. Couclelis introduced additional spaces, which she united into three groups : pre-conceptual (space of experience, sensomotor, behavioural, socio-economic), conceptual (physical and mathematical) and superconceptual (a symbolic one)[6].

On the example of publications of H. Couclelis we see an attempt (though rather clumsy) to withdraw a pure naturalistic understanding of space and to connect (to bridge) a naturalistic understanding of space with a culturological one by means of inclusion a symbolic space into a circle of spaces.

Now, after a publication of M. Heidegger's works, the phenomenological foundations of the idea of space cannot be ignored. The idea of space surpassing through the consciousness of a phenomenologist will be perceived in quite a different way. This is already not a physical space and not the numerous " departmental " spaces (separated from the physical one), which the person perceives as if from an external point, but an existential space, constituted by human consciousness. The naturalistic approach, on the contrary, tries to clear up a " spatial world picture " from " distortions ", obliged to consciousness.

It is possible to generalise, that geographers drift from the naturalistic to the culture-laden vision of space, that is connected with the gradual adoption of ideas which arrive mainly from phenomenology. A serious work in overcoming a purely physical understanding of space and in making theoretical foundations of human spatiality has been done in geography recently — in the 1980s and 1990s[7].

J.N. Entrikin emphasised the fundamental bipolarity (" betweenness ") of human consciousness which cannot but to oscillate between a " relatively subjective " and a " relatively objective " position. In the former case we are part of place (a centred cognitive view), in the later we must leave the place as a central point (a decentered cognitive view)[8]. This fundamental bipolarity of human consciousness will be important for my discourse of spatial notions in geography. This discourse is inspired by the works by Martin Heidegger and their interpretations in Russia.

5. H. Couclelis, N. Gale, " Space and spaces ", *Geografiska annaler*, 68B, n° 1 (1986).

6. H. Couclelis, " Location, place, region, and space ", in R. Abler *et al.* (eds), *Geography's inner worlds*, New Brunswick (N.J.), 1992, 215-233.

7. E. Relph, " Geographical experiences and Being-in-the-World : the phenomenological origins of geography ", in D. Seamon, R. Mugerauer (eds), *Dwelling, place and environment*, Dordrecht, Boston, 1985 ; T.R. Schatzki, " Spatial ontology and explanation ", *Annals of the Association of American Geographers*, 81, n° 4 (1991), 650-670.

8. J.N. Entrikin, *The betweenness of place. Towards a geography of modernity*, Baltimore, 1991.

THE GENERALISED CONCEPTION OF SPATIALITY

I shall try to put forward a generalised conception of spatiality in geography. It is built on overcoming a naturalistic understanding of space. The thing is that a naturalistic cognition is a necessary one which provides a progress of knowledge but not the only one.

I proceed from the assumption that the idea of spatiality or space should be comprehended not as a notion taken separately, but in connection with those spatial notions, with which a geographer actively deals — in particular, with " place ", " territory " and " region ". The idea of space should be discussed in two plans of becoming — in a logical and in a historical one. In the latter case it is necessary to reveal the mythopoetical roots of spatial ideas and to compare mythological and post mythological understanding of space. My position in this question is following : in order to realise spatiality as it is, we need for the time of discourse to give up the burden of those layers of " knowledge ", that influence our thinking, and to bring to light previously what kind of space the humans of the pre-scientific epoch could deal with. Thus, I suppose, the mythopoetical base is sure to be manifested in the dialectics of geographical concepts. The fact is that even the most abstract concepts of science should have their mythopoetical roots. The European science since the 17th century in every possible way was anxious to erase inside its own fundamentation the world of fiction and myths that at once generated the rational science of the modern type. In this article I pay attention to the roots of modern " spatial notions ", to their deep mythopoetical layers.

SPACE AND PLACE

Yi-Fu Tuan's remarkable merit consists in the following : he, the first in geography, brought together these two concepts and united them in one thinkable pair, though failed (in my opinion) to reveal their generic and logical linkage.

If to proceed from the naturalistic attitude of consciousness, geography is the description of the Earth as a single and a whole Physical body. Thus, no wonder that " the real geographical space " is seen as the only existing three-dimensional one. Under naturalistic attitude of consciousness we act in the frameworks of thought that accepts space as a thing, to say more exact, as a specific physical attribute of a thing = the technological, operational world-space. In this case our interpretation of the world begins from the entities, found " around us " instead of the phenomenon of World as a whole (with a human being as a inalienable part of it).

It begins with a corporeal thing (*res corporea*) which is characterised by substantiality. We easily isolate substantial property of extension (length, width, height), whereas other characteristics are separated from this

(" spatial ") entity. Thus, the being of the corporeal thing looks as being caused by its extension.

If we understand space in such a way, then place appears to be a location in relation to other locations (that is, a simple location in space). Such an understanding of space can completely satisfy scientists, because it emerges as a convenient pattern for the designation of the unity of diversity. In classical geography the Earth space is perceived as an already found, a taken-for-granted.

A culturological attitude of consciousness is connected not with a perception of world as the entities-in-the world but with phenomenon of the world as a whole. The idea of space (as well as of time) is taken not from things but from itself. Let us discuss how it is possible to understand space " from itself ".

Space is always connected with light, they conjugate, creating and presupposing each other. Before seeing, everything for a person is invisible, closed and even deathly-still. Only when we open the eyes, we feel that the abyss sweeps open in front of us. The subjects of sight are given to us from the outside, but it depends on us to look or not to look. By sight we co-create space. For light the backlash and an output needed — it is impossible to see something very close (butt-joint). This backlash (*Lichtung*, by Heidegger) is regarded as space.

Light transforms the closed into the opened. Space is co-created by light and it is light. When open the eyes, a person experiences the deepest feeling of the world, which is not only outside but also inside her/himself. The cognition is a distinction, a differentiation, coming to light (standing exposed) out of the mist, haze. The world in front of light becomes differentiated, divided. Not surprisingly, that the Greek Logos was associated with light (light of knowledge).

Space unconsciously emerges as uniform (a unit, an odd), continuous and absolute, whereas the being of the mankind appears divided, split, binary (a dual) one. The world as a result of the consciousness entering has lost its initial unconscious unity both " inside " and " outside " of man. It has been split, and our dealing with binary opposition becomes inevitable. The world appeared as the binary structure. The universe splintered into two main halves — the world visible and invisible, exterior and interior, this one and that one. That is why there are classic distinctions between soul and body, matter and consciousness, animated and inanimate. And what is important — space also gained its opposition — place.

We split the primary togetherness, and this split happens in a specific structure of consciousness — a binary one.

The initial meaning of the word " place " becomes more transparent if we turn to its etymology. It does not mean that we catch one occasional meaning of a word from the vocabulary. Etymology just reminds us about essential relations of what is thought.

It seems heuristic and fruitful to pay attention to the usage of the word " place " (*miesto*) in Russian language because different languages accentuate various sides of the multifaceted phenomena. The specific experience of Russian use of this word, which has not been discussed before in special literature, is able to open new unexpected perspectives to the topic under examination.

A competent etymological source asserts that a Russian equivalent for place — *miesto* is related to words a " mark ", a " sign ". They bear a great resemblance in meaning to the foreign (Indo-European) words of the same root. For example, *mietas* in Lithuanian means a stake, *miȅts* in Latvian — a pole, a post, *methos* in Irish — a post, a frontier post, *meiðr* in Old Icelandic — a pillar, a timber. The words *methis* and *methi* in the Old-Indian language also mean a post[9].

Therefore, the connection of the word *miesto* with a certain marking of land is on hand. Places as posts or stakes, being rammed in the land, form a vertical sign. It is necessary to underscore that originally these marks informed not about the appropriation of plots of land by Man (a group) for this or that purpose, but had a holy meaning. Places expressed the link between this or that plot with the divine forces, *i.e.* the participation of the specific to the general.

An additional evidence that originally, in mythological consciousness " place " meant not every place but a holy place comes from the Hebrew. In the Genesis *makom* (place) was equivalent to *makom khadash, i.e.* a holy place. [Compare with the modern Arabic word *makam* which means a place of a holy person, or a holy grave]. But as according to Jewish religion God is everywhere, place also received a character of ubiquity, which exists everywhere where God is present. In the relation between *olam* and *makom* (space and place) there is the following implicit connection : places are everywhere in the world (of course, the divine one) and, thus, they are open, while space is hidden, latent.

M. Heidegger wrote that originally the Old-German word *Ort* (place) meant *die Spitze des Speers, i.e.* the point of a spear (a javelin). A place gathers, draws something into itself and this way preserves, and what is kept becomes cleared up[10]. Let us put attention to the semantic closeness of a spear with a stake, a picket. Throwing a spear, putting a stake, or casting a glance — in all cases we make " point " labels. That is why place so easily becomes transformed in our consciousness into a point, into something that is very small, limited.

The cartographic practice particularly promoted the propagation of such an implicit opinion. Plots of land shown on usual maps are proportional to the

---

9. M. Fasmer, *Miesto (Place). Etimologicheskiy slovar' russkogo yazyka*, vol. 2, Moscow, 1967, 607-608 (in Russian).

10. M. Heidegger, *Unterwegs zur Sprache*, Tübingen, 1965, 37. Cited in : V.A. Podoroga, *Metaphysics of landscape*, Moscow, 1993 (in Russian).

area embraced by the human sight (what is possible to grasp by the human look from a hill, a mountain, or in general from a high point) are depicted on a map in the " size " of a point. As a result, usually the visible objects become " dot objects " on a map. Compare with the journalistic *cliché* : " the point on a map " or " hot points ". The fact that place is thought (though in a latent form) as something geographically indivisible, as a point (yet Aristotle indicated : the point is something that has not parts) must be admitted as very useful. In the notion of place geography received (again, in an implicit way) a sort of the " atom of space " — analogous to the corresponding conventionally " indivisible " units in physics (a particle), biology (a gene) or linguistics (a morpheme). Studies in mythology demonstrate that the human constitution of the world begins with splitting of space, and as a result the stationary point or a central axis is found as the principle of any future orientation[11]. This point becomes a centre of the world, a basis for orientation. The primary point is an initial actualisation of the archetypal pattern. This primary point is a Sign which later acquires meaning. According to the mythological cosmogony the primary point then expands and swells. In cabalistic texts the cosmogonic process is also connected with signs[12]. God engraves signs in the heavenly sphere in order to differentiate one from another. The Hebrew verb *le galif* literally means " to cut out ". Such a distinction makes the explanation of a thing.

We see and understand the world via things, subjects. But place is not a thing, it requires Man for the storage of a word, it applies to man as a call. Things open themselves via places. Space receives its essence from places. By means of places Man marks out space.

For the constructive use of consciousness, for the splitting of what has been amalgamated, for extraction of knowledge we should pay a certain price. This price is the following : the initial unity disappears and we see the objects of content, but do not perceive consciousness as it is.

Thus the subject withdraws from the structure of being for his cognition is determined by sight and standing in the out-of-the-World point. As a result space becomes an Image. This position " outside " is a privileged position. Another position — " being-in-space " — abolishes the domination of a subject. But only in this situation a person deals with the space of life. The person in the capacity of Dasein always exists in (with) space, providing the world with space, spatializes it. Space is opened by means of transforming into the human Dwelling (that is, via place). No wonder, that place is connected with the idea of dwelling : living, dwelling is the most fundamental purpose of the appropriation of place.

Noteworthy, that in the Avesta texts a word *matana* means dwelling, home. The Uzbeks and Karakalpaks in Central Asia by the word *mitan* imply an

11. M. Eliade, *The sacred and the profane*, N.Y., 1959.
12. M. Yevzlin, *Cosmogony and ritual*, Moscow, 1993 (in Russian).

abode, a camp). In Irish *meitheal* means a traditional place-based community. All these facts show that a place is thought to be first and foremost a potential dwelling, something, accommodating Man, serving him as home.

TERRITORY AND REGION

Another pair of key geographical concepts — " territory " and " region " — completely lacks the philosophical load (as well as mythopoetical roots). Nevertheless both are of exceptional importance to geographical thought. Very often these terms are regarded as synonyms, therefore an additional discursive work is required to reveal shades, nuances of their meanings. The notions " territory " and " region ", being factually identical in volume, differ in the content and in the logic of their formation.

The identity in volume means that the area, which is regarded by us as territory in one case, may be thought as a region in the other. That is quite natural since in both cases the area is presented by the identical set of places and limited by the same border. As a certain territory and region have the same geographical name (West Siberia, for instance), they are not distinguished in practical activity and instead of two poles only one is seen. I argue that territory and region are not identical notions and therefore reduction of territory to region and vice versa seriously impoverishes the conceptual apparatus of geography. The case is that we take territories and regions as ready products of thought and do not reflect upon the logic of their formation in our consciousness.

Logics of development of these notions are diametrically opposite. A territory is a certain totality of places, integrated on a common (do not mix with a single) base. For instance, the territory of the USA is a totality of places, that have a common state belonging.

A region is something that is necessarily extracted, taken from the integrated whole in the result of partitioning (and, hence, its own integrity needs verification). The idea of " region " is connected with the procedure of cutting : *Reg* in Sanskrit means " to cut ". A Russian equivalent for " region " is *rayon* — the word borrowed from the French language. A French word *rayon* in its turn originates from another Sanskrit word *raj* — " a ray " (a ray also cuts, cleaves, pierces).

The procedure of cutting and partitioning expresses the essence of the regionalization — a research method for getting and putting in order the geographical knowledge.

Territory is always an expression of an object as a whole. When we speak about a territory, we disengage ourselves from the internal differentiation, " otherness ". On the contrary, when speaking of a region we inevitably think of it as a part of the initial whole (it does not, however, prevent us from dividing a region into parts — subregions). A territory is a primordial synthetic notion, whereas a region — an analytical one. Since at first the consciousness grasps an object as something holistic, indivisible, and only after that carries out its partitioning, we can conventionally regard " territory " as a primary entity, whereas " region " as a secondary one. It is important to note that in our consciousness frequently either territory suppresses region or, on the contrary, a region supersedes a territory. Their correlation, (" twinness ", " doubleness ") as a rule is overlooked. Instead, there is an illusion of the hierarchy of these double concepts, of subordination one by another : usually region is seen as a lower level while territory as a higher one.

The genesis of notions " territory " and " region " differs from that of the notions " space " and " place ". In their formation a subject is distinctly separated from an object : a human, an actor is situated in the outer point, she/he " cuts " territories, operates with them and provides with specific meanings. In contrast to " place ", neither " territory " nor " region " can be considered as a home of Man so long as they are not generated by a profound " insideness ", imbeddedness of a human being in the experience of her/his existence. Territory and region functionally supplement each other in the geographical study process.

SCHEME 1. MATRIX OF A GEOGRAPHER'S SPATIAL THINKING

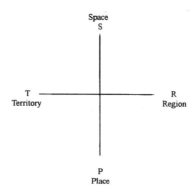

PS - axis of generation

TR - axis of establishing limit

As we have just shown, the difference between " territory " and " region " is not metrical, quantitative. They differ semantically, by logical origination. The same tells the scheme 1 : territory and region are separated from each other by the vertical axis of generation (SP). But what connects them ? The answer can be seen again on the same scheme : they are connected by the horizontal axis of establishing limits (TR). In our consciousness territory and region do have limits, though these limits can be vague and fuzzy, seen not distinctly but as a non-clear background. Much more important is the essence of the borders : in both cases — to mark a limit.

Territory and region which complement each other are important Logical concepts, Logical tools of a synthesis and analysis. As for space and place, these notions better to be attribute to phenomenological concepts.

THE TETRAD OF SPATIAL CONCEPTS

I argue that a very convenient tool which shows the logical position of notions analysed is a tetrad. The tetrad is a generic matrix enabling to distinguish a *universum*. K.G. Jung attributed the tetrad (tetraxis) to one of the principal archetypes of unconscious and considered it as a logical basis for the integral judgement in broad (philosophical) sense[13]. Specialists in biology of brain assert that an archetype (generically inherited " primeval image ") is a mental precondition and a characteristic of a cerebral function, a structure of the knowledge representation. But why precisely a tetrad ? Why a four-element matrix ?

Philosophers as well as mythologists emphasise not a quantitative, calculation function of numbers, but a logical one. Number 1 symbolises initial or pre-cosmogonic condition, determined as unpartitioned integrity. In our situation " 1 " corresponds to " space ". Number 2 symbolises an opposition and consequently distinguishness of elements. The construction of world presupposes the splitting of space, through which " the stationary point " (that is place) is found out.

Number 3 is a " minimum " number, with the help of which the description of the spatial structure becomes possible. Number 4 and a corresponding geometrical image is connected with a wholeness and a circle — a perfect form since the times of the Plato's " Timeus ". A circle was used for the designation of the light of the primary act of creation. We can find in " Timeus " by Plato the phrase that a tetraxis contains in itself all four parts of the rounded world. A tetrad symbolises parts, properties and components of the Single. This is an innate structure of the representation of knowledge. In our case the tetrad expresses the idea of spatiality in human knowledge.

The matrix of a geographer's spatial thinking (scheme 1) echoes with a matrix of knowledge by J. Schmithüsen who used it for his own examination purpose — to discriminate between 4 types of knowledge — fields of knowledge (see scheme 2)[14]. His matrix is similar to the matrix of a geographer's spatial thinking since it has the same archetypal dimensions. The poles of the

13. K.G. Jung, " Psychology and religion : West and East ", *Collected works*, vol. 11, New York, London, 1958.

14. J. Schmithüsen, *Allgemeine Synergetik*, Berlin, New York, 1976.

horizontal axis on the Schmithüsen's scheme are " general " and " special ", whereas the poles of the vertical one — " total " and " partial ". Comparing the schemes we clearly see that " space " corresponds to the general (a genus), " place " — to the special (a species, an *eidos*), " territory " — to total and region — to part.

SCHEME 2. MATRIX OF KNOWLEDGE : FOUR FIELDS OF THOUGHT[15]

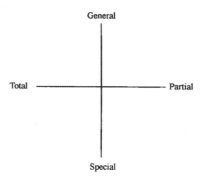

General

Total ——————————— Partial

Special

The tetrad of the spatial concepts, which is revealed and justified, is seen as a self-construed system. The graphic record of the idea of spatiality also has a spatial expression. It is given in an aphoristic form : it has no superfluous, surplus axes (aphorism : *apo* - open, distant ; *horismos* - a frontier, a border). Such a structure of spatiality does not limit the meanings of the idea of spatiality, but, quite contrary, discloses them. Thus, this graphic record is simultaneously both an expression and explanation of the structure of spatiality in the geographers' consciousness.

The axis " space - place " is the axis of the generation of meanings of the World, it reflects the links between specific and general, expresses the participation of Man to the World. The axis " space - place " in the pure state is characteristic for the mythological consciousness and suffers radical transformation in modern post-mythological consciousness (as a result, we find only shades of it).

The axis " territory - region " — is the axis of the corporal, it is connected with the usual human life, with the mortal and earthly, with the world of things. The crossing of the two mentioned axes is the cross-roads of the body and the World, the profane and the divine.

All modes of the human existence on the Earth — to build, to live, to think — determine the human experience of spatiality which has been formed in the scope of interplay of the divine and mortal, earthly and heavenly, i.e. the

15. J. Schmithüsen, *Allgemeine Synergetik*, Berlin, New York, 1976, 7-8.

Heidegger's " Geviert " (a tetrad) which he called a " topos "[16]. On the scheme 1 territory and region are shown on the horizontal axis and on the same level, as far as between them the relation of mirror opposition exists. Another pair of concepts " space " and " place " are presented on the vertical axis, a concept " place " being shown below, and space — above, because space communicates with the heavenly width, with Cosmos, that is above a human being (recollect : if we wish to see the width, it is necessary to mount). The " territory " and " region " are connected with the level of the earthly and mortal, whereas " space " and " place " — with the divine and the heavenly.

The tetrad of spatial notions is an ideal scheme which corresponds to the full becoming of notions. But in the everyday research work geographers deal with the synthetic, but not clearly distinct elementary notions. Accordingly, the matrix of a real geographer's spatial thinking receives a reduced form (see scheme 3). Geographers fail in a distinct way to distinguish between geographical notions and then clear, elementary notions are subjected to mixture and from that fusion surrogates emerge.

SCHEME 3. SURROGATES OF THE SPATIAL NOTIONS IN THE CASES WHEN THE TETRAD UNDERGOES A SEMANTIC REDUCTION

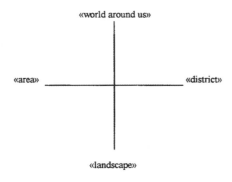

If on the " space - place " axis these two poles are not fully distinguished and articulated, they fuse in a various " weight " proportion. If the idea of place is suppressed, then we think about " the world around us " instead of space. But sometimes the idea of space become suppressed, and the balance on the vertical axis changes " in favour of place ". Then it would be more correct to call the notion appearing from this mixture " landscape ".

A similar reduction quite often occurs with another axis of spatial thought. If on the axis " territory - region " these two poles are not distinguished, either territory or region practically disappear. If we do not distinguish a territory,

16. M. Heidegger, *Being and time*, London, 1983.

then we deal actually with an " area ", if we do not notice " region ", then we deal rather with " land ".

CONCLUSION

The notion of spatiality is extremely necessary for humans to organise, put in order their knowledge about the world. The same has been done by geography : during the 20th century this discipline made a distinct drift from the concept of space as a mere container of all our terrestrial " wheres " to its deeper interpretation as a filled with meanings empirico-rational reality, in the context of which all these " wheres " acquire sense. The deeper current understanding of the concept of space had required the introduction of a new concept of spatiality. It is connected with the quaternary principle of understanding and representation of reality in its wholeness and relative static — and the archetype of tetraxis (the term used by K.G. Jung).

The simplest and most evident manifestation of the quaternarity in geography is the depiction of the four sides of horizon (north, south, east, west). But the principle of quaternarity in a spontaneous and absolutely imperceptible for geographers way was absorbed also by a conceptual (terminological) apparatus of geography. The geographical community comes to the conclusion that geographical spatiality is based on four fundamental (elementary) notions — space, place, territory and region. Space and place, on the one hand, and territory and region, on the other, present themselves 4 poles of two absolutely different but complementary and interdependent principles of spatial practice. The former principle is connected with analytico-synthetic procedures, while the latter — with the generation of meanings. By means of territory and regions we logically describe the Earth, whereas space and place are the prerequisites of the very possibility of our discourse on territories and regions. The philosophical content of the four basic geographical notions under discussion is the following : space correlates with general, place — with special, territory — with total, region — with part. The axis of meanings generation and the analytico-synthetic axis with four elementary notions-poles together form a matrix of spatiality — the generic structure of human spatial thinking.

ZUR NEUAUSRICHTUNG (BZW. NEU-BEGRÜNDUNG) DER WISSENSCHAFTLICHEN GEOGRAPHIE IM GEFOLGE DER REFORMATION DURCH MELANCHTHON, MERCATOR UND KECKERMANN. PROVIDENTIALEHRE UND GEOGRAPHIE

Manfred BÜTTNER

EINFÜHRUNG

Im Zusammenhang der Mercator-Symposien, die Anfang der neunziger Jahre begannen und die Grundlagen für eine Umbenennung der Uni Duisburg in " Mercator-Universität " bildeten, ist die Frühgeschichte der deutschen Geographie wieder näher in den Blick gekommen. Diese Frühgeschichte ist an die Namen Melanchthon, Mercator und Keckermann in besonderer Weise gebunden. Im Grunde genommen bildet sich unser Fach Geographie erst durch die Arbeiten dieser drei Gelehrten heraus, und zwar im Gefolge der Reformation. Vorher hat es zwar auch schon Werke gegeben, die man " der Sache nach " zur Geographie rechnen kann, aber es fehlte erstens noch der Name " Geographie ". Vincentius z. B. behandelt das, was wir heute zur Geographie rechnen, im Rahmen seines vorwiegend theologisch ausgerichteten Werkes " Speculum Naturale ". Und im 15. und 16. Jahrhundert betrachtete man sich als Physiker, Kosmograph oder Länderkundler (und gab seinen Werken entsprechende Titel), wenn man sich (unter anderem auch) mit dem sogenannten " geographischen Material " befaßte.

Und zweitens fehlte eine fachspezifische Zielsetzung, Forschungsmethode und vor allem ein Forschungsobjekt.

Nur, wenn die drei genannten " Essentials " (Objekt, Methode und Zielsetzung) klar definiert sind oder zumindest " der Sache nach " deutlich in den Blick kommen, kann man ja von dem Fach Geographie sprechen, das sich klar einerseits von Fächern wie Physik, Kosmographie usw. abgrenzt, aber andererseits auch eine Ergänzung zu diesen bildet.

Ich will im Folgenden herauszustellen versuchen, wie sich nach dem gegenwärtigen Stand der Forschung die Neuausrichtung und damit Begründung der wissenschaftlichen Geographie der Neuzeit im einzelnen vollzogen hat. Man kann drei Stufen herausstellen.

1. Stufe : Melanchthon. Er führt als Praeceptor Germaniae einen doppelten Übergang von dem, was man als mittelalterliche Geographie bezeichnen kann, zur "neuen" Geographie des 16. Jahrhunderts durch. Erstens löst er die "katholische" Verbindung mit der theologischen Schöpfungslehre und setzt an die Stelle die protestantische Verbindung zur (lutherischen) Providentialehre, also statt Verweis auf Gott den Schöpfer, Verweis auf Gott den Lenker und Regierer der Welt. Dadurch wird, modern gesprochen, aus einer "statischen" Geographie so etwas wie eine "dynamische".

Zweitens wendet sich Melanchthon sehr stark der Meteorologie zu und verbindet diese im Rahmen dessen, was er als Physik bezeichnet mit der Geographie, erweitert unser Fach also um das, was wir heute als Klimatologie betreiben und so bezeichnen.

2. Stufe : Mercator. Wie sich erst vor kurzem herausgestellt hat, stand Mercator mit Melanchthon in wissenschaftlichem Kontakt. Mercator geht, angeregt auch von den reformierten Naturwissenschaftlern um Danaeus einen Schritt weiter als Melanchthon. Er verbindet die Geographie nicht " nur " mit der Providentialehre (und verweist damit nicht nur auf Gott), sondern richtet diese sogar christologisch aus, indem er betont : Jegliche Geographie, Kosmographie, Physik usw. muß in erster Linie auf Jesus Christus verweisen, den Retter der Welt. Geschaffen wurde die Welt zwar von Gott, doch sein Sohn ist es, der die im Gefolge der Sünde Adams " verlorene " Welt errettet. Durch den Kreuzestod ist die Welt wieder unvergänglich, durch den Sündenfall Adams war sie vergänglich geworden. Das herauszustellen, ist die wichtigste Aufgabe jeder Naturwissenschaft.

3. Stufe : Keckermann. Er ist es, der unser Fach aus der von mir sogenannten theologischen Umklammerung oder Indienstnahme " befreit " und gleichzeitig eine fachspezifische Methode, Zielsetzung usw. durchführt. Mit Keckermann ist die im Gefolge der Reformation einsetzende Umorientierung zu einem ersten großen wissenschaftlichen Abschluß gekommen.

Aus aktuellem Anlaß, wegen des Melanchthon-Jahres 1997 werde ich mich im Folgenden stärker Melanchthon zuwenden und auf Mercator und Keckermann nur kurz hinweisen.

MELANCHTHON

Um Melanchthon, den Neubegründer des Schul- und Hochschulwesens im von der Reformation geprägten Europa richtig " verstehen " und einordnen zu können, sollte man mit seiner Grundforderung beginnen : " An den lutherischen Universitäten und Schulen sollen nur solche Fächer gelehrt werden, die in den Dienst der Doctrina Evangelica gestellt werden können. Fächer, bei denen das nicht möglich ist bzw. bislang nicht möglich war, müssen neu ausgerichtet werden... oder sind vom Fächerkanon auszuschließen ".

Wie eine in diesem Sinne umgestaltete Geographie angelegt werden sollte, führt Melanchthon aus in seiner Schrift von 1549 : *Initia Doctrinae Physicae*. Obwohl Melanchthon den Begriff " Physik " im Titel führt, handelt es sich hauptsächlich (jedenfalls aus heutiger Sicht gesehen) um eine stark meteorologisch (in Anlehnung an Aristoteles und dessen Vorstellung vom " Ersten Beweger ") ausgerichtete Geographie. Doch diese Geographie ist so stark nicht nur mit theologischen Erwägungen über die göttliche Weltregierung durchsetzt, sondern im Grunde genommen einzig und allein auf die Erläuterung dieser Weltregierung hin angelegt, daß man eigentlich von einer theologischen Schrift sprechen kann und muß. Das ist übrigens der Grund, weswegen dieses wichtige Werk bis heute weder unter Wissenschaftshistorikern noch unter Theologen angemessen bekannt und wissenschaftlich " aufbereitet " worden ist. Die Theologen wagen sich an diese Schrift nicht heran, da sie ihnen zu stark naturwissenschaftlich-geographisch ausgerichtet ist. Und ein Wissenschaftshistoriker hält sich im allgemeinen gern zurück, wenn auf weite Strecken rein theologische Begriffe und Argumentationen in den Blick kommen.

Ich betrachte es als ein Anliegen zum Melanchthon-Jahr 1997, dieses wichtige Werk endlich einmal etwas mehr bekannt zu machen... und demnächst eine Übersetzung ins Deutsche vorzunehmen.

Sehen wir uns die Schrift etwas näher an, soweit es die begrenzte Zeit auf diesem Kongreß zuläßt.

Sie besteht aus zwei Teilen. Im Vorwort und in den Eingangskapiteln (erster Teil) versucht Melanchthon, grundsätzliche Klarheit über das Verhältnis zwischen Naturlehre und Theologie (lutherischer Theologie) zu gewinnen. Im zweiten Teil bringt er das geographische " Faktenmaterial ", und zwar in der physiogeographisch-meteorologischen " Reihenfolge " " von außen nach innen ". Wenn man moderne Begriffe verwendet, läßt sich sagen : Es handelt sich um eine Geographie, die nach dem Prinzip der exogenen Dynamik angelegt ist, was ja ausgezeichnet zur aristotelischen Vorstellung vom ersten Beweger paßt.

Doch zunächst zum ersten Teil :

Von den verschiedensten Seiten ansetzend und mit immer wieder neuen Argumenten versucht Melanchthon klarzumachen, daß es für einen lutherischen Naturwissenschaftler die Hauptaufgabe sein muß, einen Weg über die Natur zu Gott zu weisen.

Unter diesem Gesichtspunkt ist folgendes auffallend : Melanchthon spricht zwar häufig von Gott, viel häufiger, als man es in einem naturwissenschaftlichen Werk vermuten würde. Der Terminus Gott (*Deus*) erscheint auf jeder Seite so oft, daß sogar das Schriftbild davon beherrscht wird. (Und ich persönlich habe Verständnis dafür, daß sich ein Wissenschaftshistoriker, vor allem, wenn er sich mit Geschichte der Naturwissenschaft befaßt, schon beim Überfliegen des Textes den Eindruck gewinnt, eigentlich gewissen muß, daß es sich wohl doch eher um ein theologisches Werk handelt, wofür er sich nicht " zuständig " fühlt).

Doch niemals ist vom Schöpfer (*Creator*) oder der Schöpfung (*Creatio*) die Rede, nicht einmal im speziellen Deus-Kapitel. Wenn Melanchthon Gott näher erläutern, " beschreiben " will, dann verwendet er Begriff wie Gubernator, Director usw., also Begriffe, die den Lenker der Welt bezeichnen, allenfalls den Werkmeister (*Opifex*), niemals den Schöpfer.

Das Gesagte macht deutlich : Es wird " nur " der aristotelische erste Beweger erreicht, der das Vorhandene lenkt, aus dem Chaos eine Maschine bildend, die perfekt funktioniert. Anders ausgedrückt : Es wird der jetzt tätige Gott erreicht, den man in der theologischen Fachsprache mit Providentia bezeichnet, nicht der biblische Schöpfergott. Mit keinem Wort weist Melanchthon darauf hin, daß der Weltregierer mit dem Schöpfergott identisch ist. Schöpfung ist in dieser Schrift kein Thema.

Der Hauptgedanke Melanchthons lautet, wie er ihn mehrfach in jeweils etwas abgewandelter Form vorträgt : " Dieses ganze herrliche Welttheater ist ein Beweis für Gott den Werkmeister (*Opifex*)… man kann den Architekten erkennen, der andauernd bei seinem Werk ist und alles lenkt. …Wir können daher in dieser Welt Gottes Fußspuren (*Vestigia*) erkennen, wenn wir Naturwissenschaft betreiben ".

Soviel als kurzer Hinweis zu Teil 1 der Schrift.

Auf den zweiten Teil brauchen wir in unserem Zusammenhang nicht näher einzugehen. Es sei lediglich darauf hingewiesen, daß die von mir sogenannte " exogene Dynamik ", nach der die Geographie Melanchthons angelegt ist, ausgezeichnet zur aristotelischen Vorstellung von ersten Beweger " paßt ", der " von oben " (also von außen) die Welt regiert. Mit der Bibel hat diese Vorstellung nichts oder nur sehr wenig zu tun. Ebenso hat sie im Grunde genommen auch relativ wenig mit der lutherischen bzw. insgesamt reformatorischen Theologie zu tun, allenfalls vielleicht mit dem " gnädigen " Gott, der jetzt und hier sozusagen beweisbar ist. Melanchthon legt nämlich eine Menge von Gottesbeweisen vor, aus denen deutlich wird, wie Gott die Welt " von oben " lenkt. Anders ausgedrückt : Melanchthon beweist mit Hilfe der stark meteorologisch-physiogeographisch ausgerichteten Geographie, daß Gott der Weltregierer ist, der alles zum Wohle des Menschen lenkt : Sein sehr oft mit verschiedenen Worten vorgetragener Gottesbeweis (ich möchte ihn den Lieblingsbeweis Melanchthons nennen) lautet etwa so : " Gott schickt zur rechten Zeit, gerade dann, wenn Menschen, aber auch Tiere und Pflanzen es nötig haben, Regen, Hitze, Wärme und Kälte. Das ist erstens ein Beweis dafür, daß es Gott als Weltregierer gibt und zweitens, daß dieser Gott es gut mit uns meint, uns zum Heil führt ".

Mit diesen Hinweisen sei zu Mercator übergeleitet.

MERCATOR

Gerhard Mercator, der nach dem gegenwärtigen Forschungsstand zwar offiziell nicht konvertiert ist, sondern auch nach seiner Übersiedlung (von Löwen nach Duisburg) Mitglied der katholischen Kirche geblieben ist, war doch in seinem ganzen Denken (vor allem auch in bezug auf die Abendmahlslehre) so stark " lutherisch " geprägt, daß man ihn " der Sache nach " zusammen mit Melanchthon und Keckermann zu den Protestanten rechnen kann. Er wurde übrigens in Flandern längere Zeit wegen " Lutherei " inhaftiert. Von daher hält sich nach wie vor die (allerdings durch nichts bewiesene) Vorstellung, daß man ihn als Glaubensflüchtling bezeichnen kann. Auf Einzelheiten sei in unserem Zusammenhang nicht eingegangen. Seine Kosmographie ist jedenfalls (wenn man so will) eigentlich " evangelischer ", protestantischer als die Geographie bzw. Physik Melanchthons.

Fragen wir, von Melanchthon herkommend : Was macht Mercator genauso wie Melanchthon, was ist bei ihm anders ?

Setzen wir mit dem zuletzt über Melanchthon Gesagten an, mit den Gottesbeweisen über das physiogeographische Faktenmaterial.

Beweise gibt es bei Mercator nicht. Für ihn steht der Glaube an erster Stelle. Obwohl man Mercator, wie auch Melanchthon, zu den strengen Aristotelikern rechnen kann, besteht in dieser Hinsicht ein großer Unterschied zwischen beiden Naturforschern. Mercator steht als Geograph, Kosmograph, Physiker usw. viel näher bei der Bibel und argumentiert viel mehr mit Zitaten aus der Bibel, vor allem auch mit dem Neuen Testament, als Melanchthon.

Es sei nur auf den einen bereits eingangs angesprochenen Punkt hingewiesen : Ähnlich wie Melanchthon ist auch Mercator der Meinung, daß man über die Natur bzw. Naturwissenschaft zu Gott gelangen kann. Ihm kommt es aber weniger auf den Regierer-Gott (die Providentia) an, sondern auf den " Gott in Christus ". Für ihn soll und darf sich ein Geograph bzw. Kosmograph nicht darauf beschränken, mit Hilfe seines Faches irgendwie zu Gott zu führen, sein Fach also " nur " theologisch indienst zu nehmen, sondern für ihn ist die Christologie das Wichtigste. Nach dem gegenwärtigen Forschungsstand dürfte Mercator hierzu von Danaeus angeregt worden sein, der als erster eine sogenannte " Christliche Physik " geschrieben hat, in der es eben um Christus, den Retter der Welt, geht, den wir Menschen über die Natur bzw. die Naturwissenschaft in den Blick bekommen, nicht um Gott den Schöpfer oder Regierer. (Diese christologisch ausgerichtete Geographie bzw. Kosmographie usw. findet sich dann in einer gewissen Unterströmung noch bei Reyher, Comenius, A.H. Francke und Zinzendorf, reicht also bis weit ins 18. Jahrhundert hinein).

Auf weitere Einzelheiten kann leider nicht eingegangen werden. Halten wir lediglich fest : In gewisser Weise steht Mercator näher als Melanchthon bei den mittelalterlichen Geographen, die sich stark von der Bibel anregen ließen, oft nichts weiter als eine Schöpfungsexegese lieferten. In bezug auf Verbindung zwischen Geographie und Doctrina Evangelica bietet er jedoch so etwas wie einen Schritt über Melanchthon hinaus, indem er nicht lediglich auf Gott, den Schöpfer und Regierer verweist, sondern auf Gott in Christus, den Retter der Welt.

Nun noch eine Art Schlußwort zu Keckermann :

KECKERMANN

Er ist in die Geschichte der Geographie als Begründer der Geographia Generalis eingegangen, wenn man so will, als Begründer der wissenschaftlichen Geographie. In seiner epochemachenden Schrift von 1616 (Systema Geographicum) entfaltet er für das Fach Geographie erstmals eine facheigene Systematik, ein fachspezifisches Objekt und eine entsprechende Forschungsmethode. Basierend auf dem, was Melanchthon und Mercator angedacht haben (unter dem Einfluß reformatorischen Denkens) greift er zwar vieles auf, was man als " rein Geographisches " bezeichnen kann, betont aber als führender Protestant (Theologe und Universalwissenschaftler), daß es an der Zeit sei, nicht nur die Geographie, sondern alle Naturwissenschaften aus ihrer theologischen Umklammerung bzw. Indienstnahme für die Theologie " herauszulösen " und theologisch völlig neutral zu betreiben. Mit einer neutralen Geographie, die ihre Ziele, Methoden, Objekte usw. selbst entfaltet, völlig unabhängig von der Theologie, ist ein viel wichtigeres theologisches Ziel zu erreichen : Durch Naturwissenschaft wird der Mensch in bezug auf die Weltbeherrschung gottähnlich, ja sogar gottgleich. Doch das wäre ein eigener Vortrag, besonders, wenn es dann um die Verantwortung geht. ...um die gottgleiche Verantwortung für die Welt[1].

KURZFASSUNG

Seit den Mercator-Symposien und der Umbenennung der Universität Duisburg in Mercator-Universität im Mercator-Jahr 1994 ist die Frühgeschichte bzw. Begründung der wissenschaftlichen Geographie in Deutschland stärker in den Blick gekommen, verstärkt durch das Interesse an Melanchthon in diesem Jahre 1997 (Melanchthon-Jahr).

---

1. Zu diesen Dingen, die uns heute mehr interessieren als Melanchthon, gab es eine sehr anregende Diskussion, sowohl in Lüttich als auch in Bonn. Einzelheiten dazu werden in den entsprechenden Publikationen von Büttner ausgeführt. Siehe dazu das Literaturverzeichnis, insbesondere auch die Habil.-Schrift Büttners aus dem Jahre 1973.

Nach dem gegenwärtigen Stand der Forschung kann man Melanchthon, Mercator und Keckermann zu den Begründern der wissenschaftlichen Geographie in Deutschland rechnen. Im Gefolge der Reformation findet im 16. Jahrhundert der Übergang von der mittelalterlichen Universalkosmographie zu den Einzelwissenschaften statt, insbesondere zur Geographie, mit einer neuen fachspezifischen Methodik und Zielsetzung in Auseinandersetzung mit der damals alle Fächer beherrschenden Theologie.

Melanchthon überführt die mit der Schöpfungslehre verbundene alte Physik in eine " neue " Geographie, in der es theologisch um die Providentia geht (um die göttliche Regierung der Welt), die im Wettergeschehen zu beweisen ist.

Mercator geht einen Schritt weiter und richtet die Geographie christologisch aus : Es ist die Aufgabe der Geographie, aufzuzeigen, daß mit Christus und seinem Kreuzestod die Welt (der Kosmos) wieder unvergänglich geworden ist.

Keckermann bündelt die Ergebnisse, löst die Geographie aus ihrer theologischen Ausrichtung und gilt daher als Begründer der neuen theologisch neutralen Geographie.

LITERATURVERZEICHNIS

Abkürzungen :

AGG-RUF *Abhandlungen zur Geschichte der Geowissenschaften und Religion / Umwelt-Forschung*, Hrsg. : Manfred Büttner, Aachen, 1988, Bochum, 1989, ff.

AQGK *Abhandlungen und Quellen zur Geschichte der Geographie und Kosmologie*, Hrsg. : Manfred Büttner, Paderborn, 1979, ff. Neue Folge Münster, 1996, ff.

GK *Geographie im Kontext*, Hrsg. : Manfred Büttner, Frankfurt, 1997, ff.

Ggrs *Geographers. Bibliographical Studies*, edited by T.W. Freeman, M. Ougthon and Ph. Pinchemel on behalf of the IGU-Commission on the History of Geographical Thought, London, 1977, ff.

GR *Geographia Religionum. Interdisziplinäre Schriftenreihe zur Religionsgeographie*, Hrsg. : Manfred Büttner u.a., Berlin, 1985, ff.

DMS *Duisburger Mercator-Studien*, Hrsg. : Hans Heinrich Blotevogel, Manfred Büttner u.a., Bochum, 1993, ff.

Phys-Theol *Physikotheologie im historischen Kontext*, Hrsg. : Manfred Büttner, Münster, 1995, ff.

A. Quellen :

Aristoteles, *Über die Welt*, Paderborn, 1952 (deutsch von P. Gohlke).

Aristoteles, *Meteorologie*, Paderborn, 1955 (deutsch von P. Gohlke).

Aristoteles, *Über den Himmeln*, Paderborn, 1958 (deutsch von P. Gohlke).

J.A. Comenius, *Physicae ad Lumen Divinum reformatae Synopsis...*, Lipsiae, 1632 (Gießen : ed. Reber, 1896).

L. Danaeus, *Physica Christiana...*, Lugduni, 1576. Weitere Auflagen, Genf, 1579-80, 1583, 1588, 1602, 1606.

L. Danaeus, *Physicae Christianae pars altera*, 1580. Genf, 1582, 1583, 1589, 1606.

L. Danaeus, *Opuscula omnia Theologica ab ipso auctore recognita et in tres classes divisa*, 1583. Genf, 1654.

B. Keckermann, *Systema S. S. theologiae*, Hannover, 1602.

B. Keckermann, *Systema compendiosum totius mathematices*, Hannover, 1603.

B. Keckermann, *Praecognita philosophiae*, Hanau, 1612.

B. Keckermann, *Systema physicum,* Hannover, 1612.

B. Keckermann, *Systema geographicum*, Hannover, 1616.

P. Melanchthon, *Johannis de Sacro Busto libellus de sphaera*, Wittenberg, 1531.

P. Melanchthon, *De astronomia et geographia*, Wittenberg, 1536.

P. Melanchthon, *Initia doctrinae physicae*, Wittenberg, 1549.

P. Melanchthon, *Loci von 1521, Loci von 1559*, Hrsg. von H. Engelland, Gütersloh, 1952.

G. Mercator, *Evangelicae historiae*, Duisburg, 1592.

G. Mercator, *Atlas sive cosmographicae meditationes*, Duisburg, 1595.

S. Reyher, *Mathesis Mosaica...,* Kiliae, 1669, ff. Mehrere Auflagen.

S. Reyher, *Mathesis Biblica...*, Lüneburg, 1714.

Vincentius Bellovacensis, *Speculum naturale*, Nürnberg, 1483.

B. Sekundärliteratur :

M. Büttner, " Die Geographia generalis vor Varenius. Geographisches Weltbild und Providentialehre ", *Erdwissenschaftliche Forschungen*, 7 (1973).

M. Büttner, " On the Significance of the Reformation for the New Orientation of Geography in Germany ", *History of Science* (1979).

M. Büttner, " Zur Konzeption der Physiogeographie bei Comenius ", *AQGK*, Bd. 1 (1979), 189-197.

M. Büttner, " Grundsätzliches zur Geschichte der Religion/Umwelt-Forschung seit der Aufklärung ", *AGG-RUF*, Bd. 4 (1990), 3-18.

M. Büttner, " Physikotheologie als theologische Disziplin, ein Gegenentwurf zur neuzeitlichen Naturwissenschaft ? Oder umgekehrt : Die neuzeitliche Naturwissenschaft, ein Gegenentwurf zur Physikotheologie ? ", *AGG-RUF*, Bd. 5 (1991), 145-158.

M. Büttner, " Samuel Reyher und die Wandlungen im geographischen Denken gegen Ende des 17. Jahrhunderts ", *AQGK*, Bd. 1 (1991), 199-215.

M. Büttner, " Neue Wege in der Mercator-Forschung ", *AGG-RUF*, Beiheft 2 (1992) (2. Aufl. 1995).

M. Büttner, " Mercators Hauptwerk, der Atlas, aus theologischer und wissenschaftshistorischer Sicht ", *DMS*, Bd. 1 (1993), 3-42.

M. Büttner, " Mercator und die Neuausrichtung der Kosmographie im 16. Jahrhundert ", *DMS*, Bd. 2 (1994), 13-49.

M. Büttner, " Samuel Reyher als praktischer Mathematiker ; ein Nachfolger Mercators ? Zum Problem der Beziehüngen zwischen Theologie und Naturwissenschaft am Vorabend der Aufklärung. Reyher, ein Kosmograph, Mathematiker, Universalwissenschaftler, Universalmathematiker ? ", *DMS*, Bd. 4 (1996), 262-280.

M. Büttner, " Samuel Reyher (1635-1714), Begründer der Stadtgeographie ", *GK*, Bd. 1 (1997), 7-13.

M. Büttner, " Philipp Melanchthon (1497-1560), Praeceptor Germaniae, und die von ihm durchgeführte Neuausrichtung der Geographie im Gefolge der Reformation. Ein Beitrag zum Melanchthon-Jahr 1997 ", *GK*, Bd. 1 (1997), 329-371.

W. Maurer, " Melanchthon und die Naturwissenschaften seiner Zeit ", in H. Grundmann (Hrsg.), *Archiv für Kulturgeschichte*, In Verbindung mit F. Wagner und A. Borst, Bd. 44, Heft 2 (1962).

W. Maurer, *Melanchthon-Studien*, Gütersloh, 1964.

W. Maurer, *Der junge Melanchthon*, Göttingen, 1967.

Ch. Ratschow, *Lutherische Dogmatik zwischen Reformation und Aufklärung*, Gütersloh, 1964.

LUCAS MALLADA AND THE CHANGING IMAGE OF SPAIN'S ENVIRONMENT

Steven L. DRIEVER

Lucas Mallada (1841-1921) was a geologist and mining engineer who is remembered and honoured for his study and cataloguing of Spain's fossil species (1875-1892) and his monumental *Explicación del Mapa Geológico de España* (Explanation of the Geological Map of Spain) (1895-1907). Always active in scientific research and writing during his long, illustrious career as a member of the *Comisión del Mapa Geológico de España* (Commission of the Geological Map of Spain) (1870-1912) and professor of palaeontology (1879-1892), he published his final study in 1914. Somehow, this family man found the time during his busy career to also write telling commentaries on Spain that examined the country's physical, economic, political, and social problems and offered reforms that would move Spain closer to the level of development characteristic of western and central Europe. In this short paper, we will focus on how Mallada convinced many Spaniards to reassess Spain's physical environment and to act to make irrigation water more available for regions subject to shortages of precipitation. These changes were vital to the largely agricultural economy of late 19th century Spain, and no one else was better prepared to explain the complex subject to a sceptical public with little understanding of modern science.

THE MYTH OF A BOUNTIFUL LAND

The entire modern history of Spain up until the last two decades of the 19th century was notable for the prevailing image of Spain as a rich and privileged country. As Spain emerged as a world power in the 16th century, the beginning of the period under question, its ascendancy was rationalized. Spain began to be portrayed as a victorious nation supported by a physical environment that was uniquely blessed compared to hot and dry Africa and cool and wet France. Thus, Spain not only had the conquistadors who could bring uncivilized nations to their knees before the altar of Christianity, but also had farmers who made her the breadbasket for many other nations.

This mythical view, usually referred to as the *Leyenda de oro* (Golden Legend), remained widespread even in the 19th century, a time when Spanish physics professors measured and recorded winds, temperatures, rainfalls, and other natural phenomena for a number of Spanish cities, and Spanish and foreign geologists constructed accurate maps of Spain's geology. In 1849, for example, Antonio Remón Zarco del Valle, the first president of the *Real Academia de las Ciencias Exactas, Físicas y Naturales* (Royal Academy of Mathematics and the Physical and Natural Sciences), insisted in a lecture to his academicians that " the conditions that Spain combines by its geographical position and its topography in support of scientific progress are and have always been numerous and exceptionally good "[1].

Despite the fact that a plethora of scientific data on natural phenomena became available to the reading public beginning in the 1850s the *Leyenda de oro* persisted. A fascinating example of this persistence is Carlos María de Castro's preliminary plan for the extension of Madrid. In the plan, Castro, an architect in the employ of Madrid's city government, included numerous tables of data on barometric pressures, temperatures, hygrometric readings, anemometric readings, cloudiness, precipitation, and evaporation collected by Manuel Rico y Sinobas, a physics professor at the *Universidad Central*, and by the *Real Observatorio astronómico* (Royal Astronomical Observatory). Nevertheless, Castro ignored much of the data in his own tables. He never commented on the extreme range between the high and low annual temperatures for 1854 (41.6° C on Aug. 22 and -10.4° C on December 30) nor on the large excess of evaporation over precipitation for the year (1845.16 mm versus 391.32 mm). He concluded his long discussion of Madrid's physical environment with the statement that : " From all the data that we have noted we can infer for Madrid, with reference to its climate, the following conclusions (...) they leave nothing to desire compared with the other points of the globe at the same latitude "[2].

Even Ricardo Macías Picavea, a Spanish geographer, writer, and publicist, although he realized that the precipitation in Spain was maldistributed and that the collection and distribution of water was Spain's " great geographic, national, indispensable, and primary problem "[3], felt compelled to laud Spain as naturally " rich, very rich " with " the most delightful sun (...) the most useful and abundant mines (...) the most lively, vigorous, and manageable race "[4].

---

1. A. Remón Zarco del Valle, " Las condiciones que la España reúne en favor de los progresos de las ciencias ", reprinted in E. and E. Garcia Camarero (eds), *La polémica de la ciencia española*, Madrid, 1970, 154.

2. C.M.[a] de Castro, *Memoria descriptiva del ante-proyecto de ensanche de Madrid*, Madrid, 1978, 53. Facsimile of 1860 publication.

3. R. Macías Picavea, *El problema nacional : Hechos, causas y remedios*, Madrid, 1899, 82.

4. *Idem*, 162-163.

Macías Picavea can be forgiven for those hyperbolas in light of Spanish society's need to stress the positive after Spain's devastating loss in the Spanish-American War of 1898. Perhaps he did not fully believe those platitudes, but certainly many Spaniards accepted them at face value.

Not long after the 1899 publication of Macías Picavea's *opus major* another Castilian writer and publicist, Julio Senador, noted unapprovingly that rural Castilians always know four things : (1) Spain is the most fertile country in the world, (2) Spain is the breadbasket of Europe, (3) they pay for everything, and (4) everything goes against the poor farmer[5]. The cocksurety of those ill-educated farmers was rooted in a dollop of erroneous figures, misinterpretations of facts, and false pride sustained by the *Leyenda de oro*. Eventually, the biggest rubes would come to realize that the *Leyenda de oro* was a myth. As the Spanish writer Andrés Sorel reported in his book on Castile, the conversation repeated over and over again in his travels through the region was " the day will come when all the countryside is abandoned. Either those who left will return, or what will happen then ? "[6].

The *Leyenda de oro* was first assaulted by a small but vocal minority of Spanish intellectuals. These individuals provided the underpinnings for Mallada's environmental representations of Spain, and thus they now will be discussed.

THE EARLY ASSAULT ON THE *LEYENDA DE ORO*

The earliest record I could find of a scientifically documented attack on the *Leyenda de oro* was an 1859 work entitled *Reseñas Geográfica, Geológica y Agrícola de España* (Geographical, Geological, and Agricultural Reviews of Spain) by Francisco Coello, Francisco de Luxán, and Agustín Pascual[7]. Coello was a geographer and cartographer, Luxán a self-taught geologist, and Pascual a forestry engineer. All three men had distinguished careers and made important contributions to this co-authored book ; however, Pascual's essay on agriculture stands out. He appears to be the first Spanish scholar to arrive at a new understanding of Spain's environment by using his own field observations and the meteorological observations collected in various Spanish cities during the 1850s. Pascual cautioned the reader that Spain is not humid, and he wrote of the Meseta Central, which comprises about half of Spain, that it is " one of the driest places of the globe, after the deserts of Africa and Asia, and (...) the farmer [there] always lives in fear of the contingency and uncertainty of the

5. J. Senador Gómez (presented by José Jiménez Lozano), *Castilla en escombros*, Palencia, 1993. Book originally published in 1915.

6. A. Sorel, *Castilla como agonía*, Madrid, 1975, 12.

7. F. Coello, F. de Luxán, A. Pascual, *Reseñas Geográfica, Geológica y Agrícola de España*, Madrid, 1859.

rains "[8]. It is likely that Mallada was familiar with this essay for the book appeared in the same year that he entered the *Escuela de Minas* in Madrid.

The second influence on the development of Mallada's environmental thesis was Manuel Fernández de Castro, who directed the *Comisión del Mapa Geológico* from 1873 until his death in 1895. Although the Commission had been formally established in 1849, it was reorganized in 1873 by a royal decree that charged it with constructing a geological map of Spain and conducting field studies of Spain's provinces. That decree also established a board of directors of six geologists, including Lucas Mallada, to assist the new director. Fernández de Castro ensured that the Commission's provincial reports, many of which are weighty volumes, emphasized causal relationships rather than mere description and addressed interregional commonalties rather than the uniqueness of each province.

Although an engineer by training, Fernández de Castro was an environmental scientist by passion. He had written a study on hurricanes when he was in Cuba (1857-1869), had presented a discourse on meteorology to the *Real Academia de Ciencias*, and he always insisted that the projects of the Commission should apply fresh geological data to the study of agriculture, mining, industry, construction, mineral waters, and underground water. The project reports, which in final or draft form existed for most provinces by 1883, were personally edited by Fernández de Castro. In his necrology for Fernández de Castro, Mallada referred to the death of his mentor as " the most painful and irreparable loss " and hailed him as a " person endowed with the finest qualities that can be united in a man of extraordinary merit "[9].

Joaquín Costa may have been an important influence on Mallada during the early years of their friendship. Scholars, beginning with the historian Ricardo del Arco, have believed up until now that there is no record of correspondence between the two men nor of one citing the other[10]. In reality, Costa, a Republican reformer and autodidact lawyer, was well-known to Mallada, a fellow Huescan, at least by 1882. They cited each other at least once, corresponded when Costa was an editor of *The Boletín de la Institución Libre de Eseñanza* (Bulletin of the Free Institute of Learning), and served together on the boards of the *Sociedad Geográfica de Madrid* (Geographic Society of Madrid) and the *Sociedad Española de Africanistas y Colonialistas* (Spanish Society of Africanists and Colonialists).

On May 28, 1880, Costa presented an address on irrigation to the *Congreso de Agricultores* (Farmers' Congress) in Madrid[11]. On that occasion, Costa

8. F. Coello, F. de Luxán, A. Pascual, *Reseñas Geográfica, Geológica y Agrícola de España*, *op. cit.*, 113.

9. L. Mallada, " Necrología : Excmo. Sr. D. Manuel Fernández de Castro ", *Revista Minera*, 46 (1985), 143.

10. L. Mallada (with a forward by R. del Arco), *Páginas selectas*, Huesca, 1925, XII and R. del Arco, *Figuras Aragonesas*, Zaragoza, 1956, 297.

11. J. Costa Martínez, *Política Hidráulica*, Madrid, 1975, 5-20.

roundly criticized the *Leyenda de oro* and urged his countrymen to acknowl-
edge they suffered from a niggardly climate and infertile soils. Whether this
lecture came to the attention of Mallada and influenced his thinking or both
men simultaneously were fashioning similar ideas on the environmental limitations of Spain we will probably never know. Certainly, Costa can be considered to have developed his environmental thesis independently of Mallada. In
his lecture, Costa cited Pascual's 1859 essay on Spanish agriculture[12]. The son
of a poor Huescan farmer, Costa also was familiar with the poverty and aridity
that typified rural Aragon beyond the valleys of the Ebro and the Gallego. Pascual had identified the extent of Spain's aridity, Fernández de Castro organized
teams of experts to study the physical geography of every Spanish province,
and Costa had publicized Spain's harsh climates and infertile soils. What distinguishes Mallada is that he was the first person to develop an elaborate thesis
on Spain's environmental limitations and to defend that proposition vigorously
throughout the 1880s.

THE MALLADIAN THESIS ON SPAIN'S ENVIRONMENTAL LIMITATIONS

The earliest indication that Mallada was fashioning a thesis on Spain's environment dates to May, 1881, when he concluded a booklet (on a new territorial
division of Spain) using the phrase *la pobreza de nuestro suelo* (the poverty of
our land)[13]. Mallada may have begun writing *Causas de la pobreza de nuestro
suelo* in June, 1881, when a massacre in Saida (near Algeria's Oran) of 137
Spanish emigrants, who apparently received little or no protection from the
French government, became a daily topic of conversation among Spaniards.
Those who asked why Spaniards emigrated to such dangerous places were told
that Spanish farmers could not compete with more efficient farms abroad,
especially the wheat farms in the United States and Russia. Mallada, however,
realized that Spain's agriculture also suffered from internal structural problems
and severe environmental limitations, things he had observed in the 1870s
while doing field work on the physical geography and geology of Huesca and
Córdoba for the *Comisión del Mapa Geológico*.

Between November, 1881, and June, 1882, Mallada published *Causas físicas y naturales de la pobreza de nuestro suelo* (Physical and Natural Causes
of the Poverty of Our Land), a series of 10 articles in *El Progreso*, an ardently
Republican daily that was quite popular among liberal intellectuals in Madrid.
Costa, then editor of the *Boletín de la Institución Libre de Enseñanza*, decided
to republish the first five newspaper articles in his own journal between January 6 and April 16, 1882. Because articles 6-10 were exclusively about Huesca,
they were not considered sufficient to interest the readers of the bulletin, which

12. J. Costa Martínez, *Política Hidráulica*, op. cit., 7.
13. L. Mallada, *Proyecto de una nueva división territorial de España*, Madrid, 1881, 30.

circulated to more than 600 members of the *Institución Libre de Enseñanza*, to scientific associations, and to editors of similar periodicals.

The *El Progreso* series alone persuaded Eduardo Saavedra, a distinguished engineer and president of the *Sociedad Geográfica de Madrid*, to invite Mallada to lecture on his new understanding of Spain's environment. The importance of the invitation cannot be overstated. At this time, the *Sociedad Geográfica* was one of the half dozen most important cultural institutions in Madrid, and the Society played an active, although unofficial, role in the forging of reforms later adopted by the liberal, coalition government. Mallada's lectures over the first five articles took place on February 7 and March 21, 1882.

The lectures covered five topics : (I) signs of Spanish decadence, (II) meteorological and orographic limitations, (III) geological limitations, (IV) the need to reverse deforestation, and (V) defects of the Spanish race. In the first part, Mallada drew on his travels along the back roads and trails, shocking his urbane audience with tales of hunger and malnutrition in rural Spain. In the second part, Mallada took Pascual's concept of Spain's aridity and refined it by comparing recent precipitation data for Spain and similar data for the rest of Europe that had been compiled by Alexander Keith Johnston, a Scottish geographer, and Achille Delesse, a French geologist. Mallada also appears to be the first Spaniard to make use of comparisons of the average elevations of most European countries that had been first presented by Gustav Leipoldt, a German geographer[14]. Mallada concluded that Spain is not only too dry, but also too mountainous. He further maintained that 90 percent of Spain beyond the humid Cantabrian region receives much less precipitation than needed.

Part III was based on the field studies and reports of the *Comisión del Mapa*. Although the terminology is sometimes dated[15], Mallada's general discussion of the geology of Spain remains highly readable today. At the end of this section Mallada summarized the meteorological, orographic, and geologic limitations in a table on Spain's land in relation to its agricultural productivity. In that table 80 percent of the land is held to be of moderate to low productivity because of all three limiting factors. Another 10 percent is considered unproductive because of the predominance of barren rock. Only 10 percent of the land is in accord with the *Leyenda de oro* ; that is " [it] makes us presume that we have been born in a privileged country "[16]. This classification has almost

14. G. Leipoldt, *Über die mittlere Hohe Europa's*, Plauen i./V : F.E. Neupert, 1874 ; O. Peschel and *idem, Physische Erdkunde*, Leipzig, 1879 (2 vols). The elevations were published in a table in vol. 1, 422-423.

15. For example, " micacita " (*mica-esquisto*), " cordillera carpeto-vetónica " (Cordillera Central), etc.

16. L. Mallada, " Causas de la pobreza de nuestro suelo ", *Boletín de la Sociedad Geográfica de Madrid,* 12 (1882), 105.

become a standard reference in Spanish scholarship after it was republished in 1890 in Mallada's book *Los males de la patria* (The Ills of the Fatherland).

In the discussion of deforestation in part IV, Mallada states that dwellers of treeless regions are subject to dry soils that, in turn, cause a " dryness " of the spirit. Although he sounds strongly deterministic to modern ears, Mallada was reflecting prevailing scientific views at the time that the lack of trees was a cause rather than an effect of such things as soil desiccation, low atmospheric humidity, and insufficient rainfall. The majority of the individuals in the audience of the *Sociedad Geográfica* opposed Mallada's conclusions not so much because they opposed environmental determinism, but because they did not want too share Mallada's pessimistic view of life in rural Spain. However, other intellectuals in Spain were not so reticent about embracing the pessimism that resulted from such reductionist naturalism. The leading figures of the Generation of 98 — Antonio Machado, Ramiro de Maeztu, Miguel de Unamuno, among others — sometimes borrowed Mallada's words and phrases (¡ *Bárbaros paises de rudos moradores* !, *la sequedad de espíritu*, etc.), almost as if they were perfectly shape nuggets of gold that could not benefit from refining[17].

Comments at the end of his lecture on February 7 convinced Mallada to address " habits of laziness " as a cause of Spain's backwardness in his follow-up lecture on March 21. Mallada's criticisms of his countrymen were original, but reflected a trend, begun in early 19th century Germany, to consider Latin peoples inferior to northern Europeans. The diffusion of social Darwinism to Spain in the late 1870s merely reinforced the tendency to praise the work ethic and expanding industrialism of the northern countries and to disparage the daily habits and traditionalism of the Mediterranean peoples. This cultural determinism, of course, was at variance with the environmental determinism that Mallada had carefully constructed in parts I-IV and served to soften the latter somewhat.

The debate over Mallada's lectures at the *Sociedad Geográfica* started on March 21 and continued on April 4, May 16, and June 6[18]. Most of the individuals in the audience reacted quite unfavourably to Mallada's environmental thesis. Federico de Botella, Chief Mining Engineer and the last surviving member of the core group that started the *Comisión del Mapa Geológico* in 1849, vehemently criticized the thesis that rural Spaniards lived in unremittingly harsh conditions. No doubt like many other members of the *Sociedad Geográfica*, he was overly optimistic about the agricultural productivity of Spain's highland regions and was nearly as bullish about people's ability to

17. For example, see José Tudela, " El primer escrito de Machado sobre Soria ", *Celtiberia*, 31 (1961), 65-72.
18. On the 16th, Mallada also presented his articles 6-10 on Huesca.

manage crises and to improve the environment through the planned use of technologies such as new dams and irrigation canals.

Other critics were as sanguine as Botella. Coello, by this time an honorary president-for-life of the Society, rejected the idea that the infertility of the land forced the emigration of Spaniards, suggesting instead, in Panglossian fashion, that the exodus was due to " the adventurous spirit of our race "[19]. Even Martín Ferreiro, a cartographer and general secretary of the Society, who began his comments with praise for Mallada's patriotism ended by avowing : " No and a thousand times no ! I resist with all my force the unavoidable conclusions that are drawn, in spite of us, from the statements made in the course of this discussion "[20].

By the end of the debates Mallada's environmental thesis was still far from gaining much support. All of the key figures of the Society were adamantly opposed to his ideas. Botella went so far as to publish his rejoinder to Mallada's lectures in *El Progreso* and in the *Boletín de la Institución Libre de Enseñanza*. Curiously, while Mallada published a reply in *El Progreso*, he did not do so in the Costa's bulletin. Costa may well have refused to publish Mallada's reply on the grounds of presenting the journals' foreign readers with a balanced picture of Spain's environment. If that were Costa's intention, he erred.

There are many signs that Mallada's views swiftly entered the mainstream of opinions over the Spanish environment and agriculture. By the fall of 1882, foreign journals had translated parts of Mallada's lectures[21]. In 1883, Segismundo Moret, a liberal deputy who would become president of the *Sociedad Geográfica* (1885-1887) and one of the leaders of the liberal party, used Mallada's concepts to defend legislation for dams and canals then under consideration by the Congress of Deputies. Moret wanted a remaking of Spanish geography to resolve the serious economic problems facing the country, particularly the uncompetitiveness of Spanish agriculture, which had been exacerbated by the agricultural crisis of the 1880s. As that crisis intensified in 1886, a commission of senators issued a report that clearly indicates the extent to which Mallada's ideas had diffused through Spanish officialdom. The report reads in part : " Evident facts, before unknown, have dispelled the pleasant illusion intended for our national self-esteem, which consisted in believing that we enjoyed a very privileged land. Spain's geological formation, with its numerous mountain chains, is the reason that in Spain the proportion of land unsuitable for all kinds of cultivation is greater than in other nations (…). The

19. " Discusión acerca de la conferencia del Señor D. Lucas Mallada sobre las causas físicas y naturales de la pobreza de nuestro suelo ", *Boletín de la Sociedad Geográfica de Madrid*, 12 (1882), 274.

20. *Idem*, 13 (1882), 50-51.

21. Rafael Torres Campos, " Reseña de las tareas y estado de la Sociedad Geográfica de Madrid ", *Boletín de la Sociedad de Madrid*, 13 (1882), 313.

meteorological phenomena are not favourable either (...) and today nobody does not know that Spain is the region of Europe with the least amount of rainfall to water its fields "[22].

In 1887, the government initiated a massive investigation of the state of agriculture in Spain. The report released at the end of this inquiry reveals that Mallada's assessment of Spain's environment had gained wide acceptance, even in the *Sociedad Geográfica de Madrid*, which had submitted a written response (one of 516) lamenting Spain's infertile lands, lack of irrigation, deforestation, and sequestered farms[23]. Mallada himself was one of 36 individuals who gave oral presentations before the commission overseeing the investigation. Unfortunately, he was the last speaker scheduled for October 26, 1887, and some of the previous speakers that day had been long-winded. He therefore did not have sufficient time to present his entire paper. The only new idea he presented regarding his environmental thesis was the suggestion that the only solution to Spain's economic crisis, which to Mallada was more environmental than agricultural, was to build as many irrigation canals as possible, financed by savings made in the state's budget[24].

In 1888, Mallada republished *La pobreza de nuestro suelo* as the first chapter in *Los males de la patria*, which was serialized in *Revista Contemporánea*, then one of the best Spanish journals of science and culture, and was subsequently released as a book in May, 1890. The changes in *La pobreza de nuestro suelo* made in that chapter were relatively minor, reflecting factual updating and polishing of the writing, except that the discussion of the Spanish race was gathered into its own chapter. Whether intentional or not, the separation makes a strong statement that the limitations of Spain's land exist independently of any supposed defects in the character of Spaniards. By re-examining Spain's environmental restrictions in *Los males de la patria*, Mallada introduced his thesis to a wider audience, in Spain and beyond.

CONCLUSIONS

It is difficult to imagine that, after 1890, an educated, intellectually-inclined Spaniard had not at least heard of Mallada's *Los males de la patria* and had some inkling of Mallada's ideas on the poverty of Spain's land. Perhaps this is why Mallada, after that date, did not re-examine his thesis on Spain's environmental limitations, although other scholars came forward to defend or to attack his ideas. Today, the Malladian view on the environmental reasons for Spain's

22. Joaquín Sánchez de Toca, *La crisis agraria Europea y sus remedios en España*, Madrid, 1893, 201-03.

23. *La crisis agrícola y pecuaria*, Madrid, 1887-1888 (7 vols). Especially see El Conde de Toreno (Francisco de Borja Queijo de Llano), *Contestación de la Sociedad Geográfica de Madrid*, 5 (1888), 26-34.

24. " Actas de las sesiones de la información oral ", *loc. cit.*, 6 (1888), 307-311.

economic backwardness is still widely shared among geographers and economic historians[25]. With the centenary of the Spanish-American War of 1898 and the efforts to regenerate Spain immediately following (what in Spain is still called) " the disaster ", Mallada's writings once again are reaching a wide audience. Perhaps his ideas, never completely free from controversy, once again will stir debate over the true nature of the geography of Spain, which, no doubt, has been remade considerably over the last century.

25. F.J. Ayala-Carcedo, " Medio físico y desarrollo en España. Una perspectiva histórica ", *Boletín Geológico y Minero,* 108/2 (1997), 189-216.

THE INFLUENCE OF SCIENTISTS ON PUBLIC POLICY REGARDING ROTTNEST ISLAND AND GARDEN ISLAND, WESTERN AUSTRALIA : TWO COMPARABLE ISLANDS - TWO CONTRASTING APPROACHES

Marion HERCOCK

INTRODUCTION

Where certain environments, such as island reserves, have been used for research purposes, scientists as a professional group, have contributed to the evolution of public policy. Rottnest Island and Garden Island are of interest to earth and life scientists owing to their biota and physical characteristics. Some of these features are common to the islands, others are unique to one. The two islands are the largest of a coastal chain located off the metropolitan coast of Perth, Western Australia. They enjoy a Mediterranean climate, and are formed of limestone, covered with wind-blown sands of Holocene origin, and are both used for recreation and nature conservation. The 1.800 hectare Rottnest has salt lakes and swamps, and has an east-west orientation, 18 kilometres off the mainland. In contrast, the 1.200 hectare Garden Island runs north-south, parallel to the mainland, and is only two kilometres away from the coast (fig. 1). Neither island was inhabited by Aboriginal people until Rottnest was used as a prison for natives last century. The islands are of particular interest to life-scientists as they are within easy reach of the metropolitan area, and offer an experimental " laboratory ". In addition, the fauna exhibit subtle physiological differences from mainland populations and the island populations are large enough to have developed variations within the island population[1]. The species of greatest interest are two marsupials — the quokka, a small native wallaby (*Setonix brachyurus*) which occurs on Rottnest, but not on Garden Island, where a different wallaby, the tammar (*Macropus eugenii*) is resident. Both

1. A.R. Main, " Rottnest Island as a location for biological studies ", *J. Roy. Soc. W. Aust.*, 42, n° 3 (1959), 66-67 ; H. Waring, Introduction to special issue, " Rottnest Island : The Rottnest Biological Station and Scientific Research ", *J. Roy. Soc. W. Aust.*, 42, n° 3 (1959), 65-66.

FIGURE 1. LOCATION MAP ROTTNEST ISLAND AND GARDEN ISLAND,
WESTERN AUSTRALIA

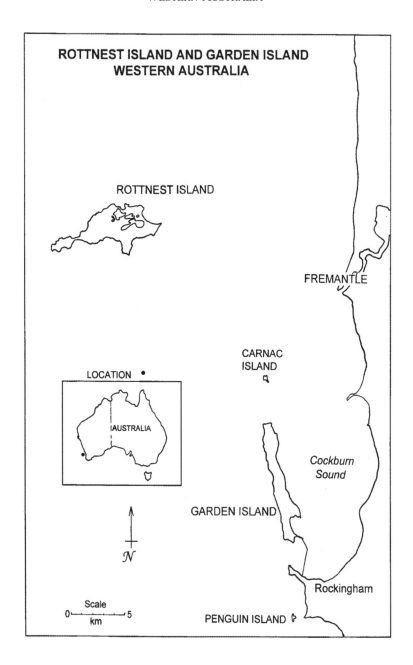

islands are used for nature conservation, partly as a result of their scientific value, but each is administered by different government agencies for different primary purposes. Rottnest is a holiday resort and nature reserve for public recreation, and Garden Island houses a naval base, HMAS *Stirling*, and is used for defence.

Although the islands have much in common physically, being within 20 kilometres of each other, they have a contrasting vegetation cover due to rainfall differences, the impact of different land-uses in the 19th century, and grazing by native marsupials. Up to 90 years ago, Rottnest was a prison for Aborigines and a summer resort for the British governors of the Colony of Western Australia. In contrast, the taller and denser vegetation on Garden Island restricted development. As a result of burning, clearing for agriculture and firewood collection, Rottnest is now largely covered by a garrigue-like scrub-heath dominated by the prickle-lily (*Acanthocarpus preissii*), and its original pine and tea-tree (*Callitris-Melaleuca*) low forest survives only in patches. Attempts to reafforest the island have not been totally successful, owing to the proliferation of grazing quokkas (the population in 1983 was estimated at 8.000 to 12.000 animals[2]), which are partial to young saplings and capable of ring-barking a tree[3]. In contrast, Garden Island has retained much of its original vegetation cover and has not suffered from an over-population of its grazing marsupial, as the population of tammars was no more than 2.000 animals in the 1980s[4].

The following discussion will examine the contrasting role that has been played by scientists, and their influence on policy concerning the two islands at different times in the last 50 years. The contrasting approaches of the scientists in that period reflect older, historical contrasts between dualism and holism, and show the impact of environmentalism upon the management of public lands.

1. The legal structures and managerial arrangements for Rottnest and Garden Islands, and the role of scientists today.

2. The contribution made by scientists to biological conservation : why scientists are in policy making positions.

3. Comparisons and contrasts between the roles of scientists on Rottnest and on Garden — what does this show ?

4. Conclusion — the analysis tells us something about :

2. Rottnest Island Management Planning Group, *Rottnest Island Draft Management Plan*, vol. 1, Feb. 1985, Perth, 1985, 60.

3. T. Sten, " An Appreciation of Progress in Reafforestation of Rottnest Island : 1954-1974 ", draft MS, records of the *Terrestrial Ranger*, Rottnest Island Authority c. 1974.

4. D.T. Bell, J.C. Moredoundt, W.A.T. Loneragan, " Grazing pressure by the tammar (*Macropus eugenii* Desm.) on the vegetation of Garden Island, Western Australia, and the potential impact on food reserves of a controlled burning regime ", *J. Roy. Soc. W. Aust.*, 69, n° 3 (1987), 89-94.

(i) the islands — bio-physical : the effect on the vegetation ;

(ii) science — the narrow approach of the specialists versus the wider approach of the generalists ;

(iii) public policy — the results produced by different political institutions.

THE LEGAL STRUCTURES AND MANAGERIAL ARRANGEMENTS FOR ROTTNEST AND GARDEN ISLANDS, AND THE ROLE OF SCIENTISTS

(1) Rottnest Island is State Government property reserved for public recreation since 1917, and over the last five years received an annual average of 291 600 visitors via the ferry service, air service and private boats[5]. The island is vested in the Rottnest Island Authority, a statutory body set up by the *Rottnest Island Authority Act 1987*. The functions of the Authority, set out in the Act clearly delineate the hierarchy of duty and public service.

" The control and management of the Island is vested in the Authority for the purpose of enabling it :

(a) to provide and operate recreation and holiday facilities on the Island ;

(b) to protect the flora and fauna of the Island ; and

(c) to maintain and protect the natural environment and the man-made resources of the Island, and to the extent that the Authority's resources allow, repair its natural environment ".

The Rottnest Island Authority is responsible for the implementation of the Act and consists of a board of management (six members) and an administrative agency. The agency carries out the day-to-day operations of the Authority under the management of the Chief Executive Officer who is subject to the direction of the Authority board. This management body determines policy and administers the agency through the office of the Chief Executive. The Authority is advised on specialist policy areas by four committees made up of experts, predominantly from government agencies and tertiary institutions. The committees are : the Kingston Advisory Committee, the Environmental and Research Advisory Committee, Heritage Buildings and Sites Committee, and the Moorings Advisory Committee (fig. 2). Members (past and current) of the Authority board, and agency staff also serve on the advisory committees[6]. Overlapping membership of the other committees and bodies associated with organisations, other than the Rottnest Island Authority, is normal.

Scientists play a role in developing policy through their membership of the Environmental and Research Advisory Committee and the Rottnest Island Foundation. This latter body functions as a channel for research funding identified by the former. The Foundation is made up of past members of the Rott-

---

5. Rottnest Island Authority, *Annual Report 1995/96*, Fremantle, RIA, 1996, 9.

6. Rottnest Island Authority, *Annual Report 1993/94*, Fremantle, RIA, 1994 ; Government of Western Australia, *Rottnest Island Draft Management Plan, Sept. 1995*, Perth, 1995.

nest Island Authority and its advisory committees, as well as invited members of the public. As scientists have rendered advice to the Authority since its inception in 1917, and occupied committee positions since the 1940s, they form a significant influence on policy concerning the island, but also have their own agenda regarding research. Biological research on Rottnest Island has been important to the scientific community and tertiary educators in Western Australia for the past fifty years. Scientists regard islands as useful field laboratories, because of their isolation, biota, and their being of some analogy to mainland nature reserves[7]. Rottnest was described by Professor H.W. Waring of the University of Western Australia's (UWA) Zoology Department, as a " gift from heaven ", which with a biological field station would provide opportunities for education and research[8]. Following the establishment of the field station in 1953, scientists were in a position to advance their interests through public policy. However, it is worth noting that neither the Rottnest Island Authority Act or its delegated legislation, the 1988 Regulations, make any reference to scientific research or its conduct.

FIGURE 2. ROTTNEST ISLAND MANAGEMENT STRUCTURE

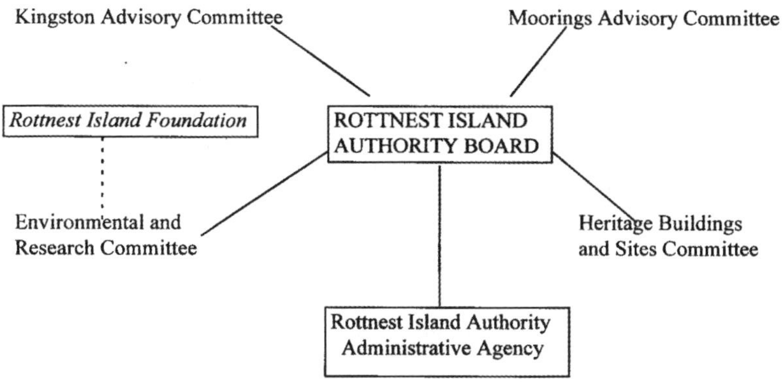

(2) Garden Island is used for recreation, but only within daylight and restricted to designated picnic areas[9]. Unlike Rottnest, it has no holiday accommodation and is only accessible by private boat. Civilians require permission from the Royal Australian Navy to use the causeway linking the island

7. A.R. Main, " Rottnest Island as a location for biological studies ", *op. cit.* ; A.R. Main, " The occurrence of Macropididae on islands and its climatic and ecological implications ", *J. Roy. Soc. W. Aust.*, 44, n° 3 (1961), 84-89 ; A.R. Main, M. Yardev, " Conservation of macropods in reserves in Western Australia ", *Biol Conserv.*, 3, n° 2 (1971), 124-132.

8. H. Waring, Introduction to special issue, " Rottnest Island : The Rottnest Biological Station and Scientific Research ", *op. cit.,* 1959.

9. Navy Public Affairs (WA), *Welcome Aboard HMAS Stirling Fleet Base West*, Rockingham, c.1995.

to the mainland. Public access is restricted as the primary purpose of Garden Island is to play a part in the defence requirements of the Commonwealth Government. Thus, the ultimate responsibility for Garden Island rests with the Commonwealth Government, owing to its powers and responsibility for defence as defined in s. 51 (vi) of *The Constitution*. As Garden Island is federal property (since 1915), it is subject to all legislation applicable to Commonwealth land and property. Management of the island is the responsibility of the Department of Defence (Navy), operating out of HMAS *Stirling*. Although the island has no legal status as a nature reserve, it is the subject of an agreement (hereafter referred to as the 1979 Agreement) between the Commonwealth and the State of Western Australia which guarantees State involvement " in scientific and environmental protection activities on the island to ensure the adequate management of the ecology of the island "[10].

As this agreement derives from the provisions of the Commonwealth *National Parks and Wildlife Act 1975* ; and the Western Australian *Wildlife Conservation Act 1950* and the *National Parks Authority Act 1976* (replaced by the *Conservation and Land Management Act 1984*), there is a clear intent to treat much of the island as a wildlife reserve with public access for recreation and scientific research. The 1979 Agreement provides for joint State-Commonwealth management of the areas outside the naval facilities zones, and towards this end, established a joint State-Commonwealth advisory body to assist the Royal Australian Navy in managing the island ecology and public access. This body is the Garden Island Environmental Advisory Committee (GIEAC).

The legislation governing Garden Island is in the following order of hierarchy. As the land is owned by the Commonwealth, its laws prevail over State law.

(i) - *Defence Act 1903* (Cwlth) ;

(ii) - *Crimes Act 1914* (Cwlth) ;

(iii) - *National Parks and Wildlife Act 1975* (Cwlth) ;

     Wildlife Conservation Act 1950 (WA) ; and

     Conservation and Land Management Act 1976 (WA).

The Defence (Public Areas) By-Laws 1987 constitute the secondary or delegated legislation to the most important Act, the *Defence Act 1903*. Under s.116ZT of the Act additional by-laws may be made " providing for the collection of specimens and the pursuit of research in public areas for scientific purposes ".

The membership of the GIEAC (fig. 3) mirrors the hierarchy of the legislation, in that the committee is chaired by the Commanding Officer of HMAS

---

10. Public Service Board of Western Australia, Internal File n° 223902, *Statement of Arrangements between the Commonwealth of Australia and the State Government of Western Australia concerning the control of public access and environmental management*, Garden Island, Western Australia, Perth, 20 Feb. 1979.

Stirling. The other two permanent representatives are from the Commonwealth government, and the State government. In 1995, the Chief Executive Officer of the National Trust (WA) was co-opted to the GIEAC, on a temporary basis, as a community representative. The GIEAC is assisted by the Environmental Manager and the Ranger, these being responsible for the administrative of environmental management policy on the island. The Manager is a Commonwealth employee and the Ranger, seconded from the State Department of Conservation and Land Management (CALM) to the Navy. The Commonwealth representative on the GIEAC is an ecologist and senior research scientist from the Commonwealth Scientific and Industrial Research Organisation (CSIRO) and the State representative is the Swan District Manager of CALM[11]. The GIEAC acts as an advisory group to the Commanding Officer, with whom responsibility for the management of the island rests. Scientific interests appear well represented although CALM, as the administrator of recreation in terrestrial and marine reserves and parks, can also represent recreational interests.

FIGURE 3. GARDEN ISLAND MANAGEMENT STRUCTURE

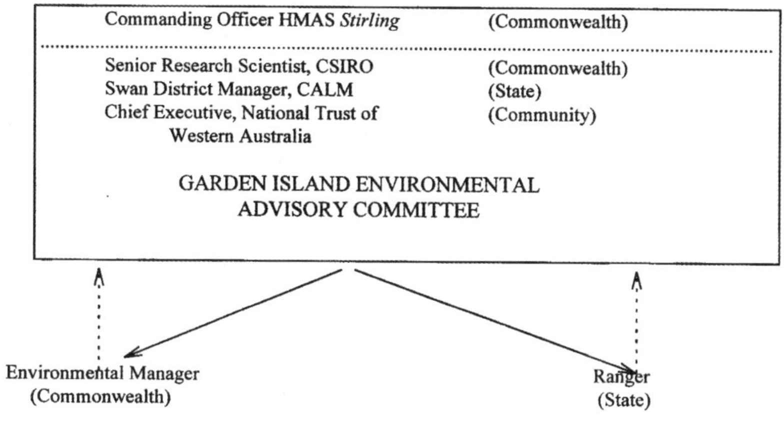

THE CONTRIBUTION MADE BY SCIENTISTS TO BIOLOGICAL CONSERVATION :
WHY SCIENTISTS ARE IN POLICY MAKING POSITIONS

(1) Biological and environmental research is essential for nature conservation, from wildlife management in the field to top-level policy-making. The literature on the topic produced by biologists, zoologists, botanists, ecologists,

11. Garden Island Environmental Advisory Committee, *HMAS Stirling and Garden Island Opportunities for Environmental Research Projects 1996-1997*, Rockingham, 1996.

and biogeographers, is vast. Regional examples may be found in edited volumes by Saunders[12] *et al.*[13] ; and the journals *Pacific Conservation Biology* and *Biological Conservation*. Today, in the conservation and environmental policy making arenas, specialist scientific skills are integrated with those from sociology, planning, economics, law and management.

The specialist has particular knowledge and policy makers turn to specialists for advice in conditions of uncertainty, and/or because of legal requirements to do so. Owing to the increasingly technical and complex nature of public governance and the specialisation and professionalisation of the bureaucracy, scientists are occupying policy making positions as consultants, employees and advisers[14]. The contribution of specialists to policy formulation has been termed " the policy role of the knowledge elite "[15] and is reflected in the deference given to scientists by policy makers and leaders. However, one must recognise that political and bureaucratic leaders are likely to remain in control of the decision-making process and will use expert opinion selectively to justify decisions. (See Walker on how the views of experts may be sought but ignored where they do not support a desired course of action[16]).

(2) Rottnest Island

The Rottnest Biological Research Station was established in 1953. The interest of UWA zoology staff, especially that of the then departmental head, Professor H.W. Waring, in continuing the student camps and research started in the 1940s, led to the proposal for a field station on the island. A portion of the land was leased from the Commonwealth (which had owned it as a meteorological and signal station during the Second World War) to the WA Department of Fisheries and Wildlife for accommodating research activities and researchers. The lease was accompanied by the establishment of a management committee — the Rottnest Biological Station Committee, which was made up of the Minister for Fisheries, who was also Chairman of the Rottnest Island Board at the time, a scientist from CSIRO Fisheries Division, the head of the Zoology Department UWA, a scientist from the CSIRO's Wildlife Survey section, a representative from the Fisheries Department and a scientist from the

12. Denis Saunders is a CSIRO scientist, and currently represents the Commonwealth Government on the GIEAEAC.

13. D. Saunders, G.W. Arnold, A. Burbridge, A.J.M. Hopkins (eds), *Nature Conservation : The Role of Remnants of Native Vegetation*, Chipping Norton, 1987 ; D. Saunders, R. Hobbs (eds), *Nature Conservation. 2 : The Role of Corridors*, Chipping Norton, 1991 ; D. Saunders, R. Hobbs, P. Ehrlich (eds), *Nature Conservation. 3 : Reconstruction of Fragmented Ecosystems - Global and Regional Perspectives*, Chipping Norton, 1993.

14. D. Collingridge, C. Reeve, *Science Speaks to Power : The Role of Experts in Policy Making*, London, 1986 ; P.M. Haas, " Introduction : epistemic communities and international policy coordination ", *International Organization*, 46, n° 1, 1-35.

15. D. Nelkin, " Scientific knowledge, public policy, and democracy ", *Knowledge Creation, Diffusion, Utilization*, 1 (1979), 107, cited by P.M. Haas, 1992.

16. K.J. Walker, *The Political Economy of Environmental Policy : An Australian Introduction*, Kensington, 1994.

W.A. Museum[17]. This committee later evolved into the Research Advisory Committee. The scientific view of Rottnest and its wildlife, particularly the quokka, as a resource for research was expressed by UWA zoologist, later Emeritus Professor, A.R. (Bert) Main : " we need rather idealised animals which are amenable to morphological, physiological and experimental analysis. Fortunately, Rottnest contains four vertebrate species which, in varying degrees, fulfil the ideal, these are the quokka (*Setonix brachyurus*) and the frogs *Heleioporus eyrei, Hyla raniformis* and *Crinia insignifera* "[18].

The volume of zoological papers and theses that has been generated by Rottnest researchers is greater than that produced by other disciplines working on the island. Until recently, very few integrative studies of vegetation dynamics or generalist environmental research (such as Hesp *et al.*'s 1983 land capability study[19]) had been carried out, in comparison to the highly specialised studies into animal physiology and epidemiology[20]. The relationship between biological research and nature conservation on Rottnest has not been entirely symbiotic, or mutually supportive. The environmental impacts resulting from the policy influence of scientific researchers will be discussed later.

The contribution of scientists to the comprehensive environmental management plan for Rottnest produced in 1985 was the result of other political events, outside of the activities of the (then) Rottnest Island Board.

(3) Garden Island

Scientists came to occupy policy-making positions regarding Garden Island more recently than on Rottnest, and owe their involvement on the GIEAC to the events following the Commonwealth decision to build a naval base on and a causeway to the island.

The Commonwealth completed feasibility studies into the establishment of a naval support facility on Garden Island in 1967, and determined in 1969 to commence development with the construction of a causeway to the island. The original basic design criteria were defined by the Department of the Navy and the causeway was to :

(i) - meet the requirements of the Navy ;

17. E.P. Hodgkin, " History of the Rottnest Biological Station ", *J. Roy. Soc. W. Aust.*, 42, n° 3 (1959), 68-69.

18. A.R. Main, " Rottnest Island as a location for biological studies ", *op. cit.*, 67.

19. P.A. Hesp, M.R. Well, B.H.R. Ward, J.R.H. Riches, " Land Resource Survey of Rottnest Island : An Aid To Land Use Planning ", *Dept of Agriculture Bulletin*, n° 4086 (1983).

20. See S.D. Bradshaw, " Recent endocrinological research on the Rottnest Island quokka ", *J. Roy. Soc. W. Aust.*, 66, n° 2 (1983), 55-61 ; R.P. Hart, J.B. Iveson, S.D. Bradshaw , T.P. Speed, " A study of isolation procedures for multiple infections of Salmonella and Arizona in a wild marsupial, the Quokka Setonix brachyurus ", *Jour. Appl. Bact.*, 53, n° (3) (1983), 395-406 ; B. Kakulas, " The nutritional myopathy of the quokka as a model for research in muscular dystrophy ", *J. Roy. Soc. W. Aust.*, 66 n[os] 1, 2 (1983), 52-55 ; A.R. Main, " The study of nature - A seamless tapestry ", *The Royal Society of Western Australia Medallist Lecture, J. Roy. Soc. W. Aust.*, 78, n° 4 (1995), 91-98.

(ii) - be economical ;

(iii) - be compatible with the State Government's (that is, the Fremantle Port Authority's) Developmental Plan for Port Facilities in the Rockingham Area [21]; and

(iv) - have a minimum effect on the local natural environment[22].

At the same time, the Geography Department of UWA was undertaking ecological, circulation and geomorphological studies in Cockburn Sound. This research was a step towards integrated research in the Sound with the Botany and Zoology Departments of the University. The Fremantle Port Authority had sponsored a post-graduate research student in geography, Peter Waterman, to carry out hydrological studies of the Sound[23], under the supervision of geographer Dr Joseph Gentilli. In the course of his work, Waterman developed an understanding of shoreline stability, the distribution of seagrass meadows and of water movement and circulation processes in the area[24], and by 1970 had joined an environmental consultancy company — Environmental Resources of Australia (ERA)[25].

As the project proposed was a major capital work, it required the approval of the Commonwealth Parliament and was therefore subjected to the scrutiny of the Standing Committee on Public Works (PWC). The PWC is a joint House of Representatives and Senate Committee, made up of members from all political parties. It examines capital works projects costing over $2 million by calling for expert opinion from inside and outside of the bureaucracy, visiting the sites of proposed works and taking evidence from interested parties and community groups. The PWC reports to the Parliament and may suggest alternatives in the works programme as it sees the need. It provides a " second opinion " from outside the public service, which often results in savings or improvements in design[26].

The Committee met in 1970 to consider the causeway design criteria, and heard from 26 witnesses : from the Commonwealth Works Department, the Navy, the State government, local government, a union representative, recreational interests, former island lessees and five earth and life scientists. The evidence of those scientists and other expert evidence produced by the consult-

21. Fremantle Port Authority, *Development of Outer Harbour, Port of Fremantle*, Fremantle, 1966, vols 1, 2.

22. Commonwealth Department of Works, " Naval Support Facility Western Australia, Point Peron-Garden Island Causeway Project ", *Statement of Evidence to Parliamentary Standing Committee on Public Works* (Melbourne, Commonwealth Department of Works, c. 1970).

23. R.G. Chittleborough, *Conservation of Cockburn Sound (Western Australia) : A Case Study*, Parkville, Victoria, 1970.

24. P. Waterman, *Southern Flats case study area, report for Fremantle Port Authority*, 1969 (unpublished ms).

25. Mr Peter Waterman, founding Director, Environmental Resources of Australian ; Director, Environmental Management Services Pty Ltd, personal communication, 23 July 1997.

26. D. Solomon, *Australia's Government and Parliament*, 4th edition, Melbourne, 1978, 50.

ants (ERA) hired by the Commonwealth Department of Works (CwlthDW), led to a change in the original design proposed by the Fremantle Port Authority — from a solid causeway to create port facilities — to a causeway with two bridged openings to allow for the exchange of water between Cockburn Sound and the open ocean " to an extent assessed as necessary to minimise any effect the causeway may have on natural environmental conditions in the area "[27].

In addition, the considerations of the Committee led the Commonwealth Department of Works to continue to employ the ERA consultants. This pioneer[28] environmental company had conducted the preliminary geomorphological, hydrological and ecological studies of Cockburn Sound in preparation for the causeway. The studies to be integrated with the engineering investigations were carried out by ERA's founding directors, zoologist Timothy Meagher, analytical chemist Ron Sheen and geographer Peter Waterman[29].

The 1972 report of the PWC on the construction of the naval support facility (HMAS *Stirling*) followed the same consultative enquiry, hearing evidence from government departments, scientists, recreational and conservation interests[30]. Again, changes were made to the proposed project and matters first raised in the 1970 report were given greater consideration — public access to Garden Island, environmental protection and items of historical interest — were dealt with by a State Government Working Group of experts, chaired by UWA zoologist Bert Main. The Group made recommendations based on the input of State Government Departments, the scientific community and the public, and led to the 1979 Agreement (mentioned earlier) between the State and the Commonwealth, designed to facilitate State scientific and environmental protection interests on Garden Island — hence the establishment of the GIEAC[31].

27. Parliamentary Standing Committee on Public Works, *Report relating to the proposed construction of the Point Peron-Garden Island Causeway, (Naval Support Facility, Cockburn Sound), Western Australia*, 22nd report of 1970 of the Committee, Parliamentary Paper n° 191, Canberra, 1970.

28. ERA Pty Ltd was the first environmental company in Western Australia, and provided marine biological, hydrographical, hydrological, coastal geomorphological and environmental chemistry services.

29. Commonwealth Department of Works, c.1970 ; Sheen Laboratories, Environmental Chemistry Division, *Report on Hydrology of Cockburn Sound : Summer*, Perth, Commonwealth Department of Works, 1970 ; Environmental Resources of Australia, " Beach Morphology Kwinana-Cape Peron Foreshore, Early Summer 1970-1971 ", unpublished report, Perth : Commonwealth Department of Works, 1971 ; Environmental Resources of Australia, " The Ecology of Cockburn Sound : Summer 1970-1971 ", unpublished report, Fremantle : Fremantle Port Authority, Aug. 1971 ; Environmental Resources of Australia, " An Integration of Hydrological Investigations : Cockburn Sound, December 1971 ", unpublished report, Perth : Commonwealth Department of Works, 1972 ; Dr Joseph Gentilli, Honorary Research Fellow, Department of Geography, University of Western Australia, pers. comm., Apr. 1997 ; Waterman, pers. comm. 1997.

30. Parliamentary Standing Committee on Public Works, *Report relating to the proposed construction of a Naval Support Facility (HMAS Stirling) at Cockburn Sound, Western Australia*, 7th report of 1972 of the Committee, Parliamentary Paper n° 37, Canberra, 1972.

31. Public Service Board of Western Australia, 20 Feb. 1979.

In the years following scientists were contracted to conduct base-line studies for a " land management plan ", as very little scientific research had been
carried out since the botanical work of McArthur and Baird in the 1950s[32].
The preliminary work was started by CSIRO scientists from the Division of
Land Resources Management, who undertook an integrative ecological study
of the island for the Department of Defence's Land Management Plan for Garden Island[33]. This work contributed to the final Environmental Management
Plan which was completed by geographer Peter Waterman[34]. The plan is now
administered by the Navy and together with the GIEAC, underwrites environmental policy on Garden Island and sets the standards for the conduct of scientific research.

COMPARISONS AND CONTRASTS BETWEEN SCIENTISTS ON ROTTNEST AND ON GARDEN ISLAND — WHAT DOES THIS SHOW ?

The policies concerning the management of the two islands and how they
are used for conservation and research, owe something to scientific experts.
However, the involvement of scientists has been for different reasons and I will
now put forward explanations for the contrasts between the two groups and
island policies.

(1) The first contrast is in policy, as reflected in the legislation (primary and
delegated), the relevant management plans and other public documents. The
GIEAC's role to evaluate and supervise scientific research conducted on Garden
Island, is to ensure that research is linked to, and for the benefit of, management. Equally, the Committee " is likely to deny access for research that does
not clearly benefit understanding and management of the island's special environmental values "[35].

In stark contrast, there are no such formal requirements on Rottnest, except
for some principles outlined by the Rottnest Foundation. The most significant
difference is that Garden Island policy places scientific research in a supportive
role and integral to management, in contrast to Rottnest policy. Research on
Garden Island is carried out at the discretion of the GIEAC, under its scrutiny
and is to be of use to management. On Rottnest, policy has permitted scientists

32. W.M. McArthur, " Plant ecology of the coastal islands near Fremantle, Western
Australia ", *J. Roy. Soc. W. Aust.*, 40, n° 2 (1957), 46-64 ; A.M. Baird, " Notes on the regeneration
of vegetation of Garden Island after the 1956 fire ", *J. Roy. Soc. W. Aust.*, 41, n° 4 (1958), 102-
107.

33. Department of Defence, *Land Management Plan for Garden Island*, Canberra, Department
of Defence, 1980 ; W.M. McArthur, G.A. Bartle, *The Landforms, Soils and Vegetation as a Basis
for Management Studies on Garden Island, Western Australia*, CSIRO Land Resources Management Series n° 7, Melbourne, 1981.

34. Environmental Management Services, *HMAS Stirling, Garden Island, Western Australia.
Environmental Management Plan*, Canberra, 1993.

35. Garden Island Environmental Advisory Committee, 1996.

to operate in relative isolation, detached from management, without linking research to management needs.

The contrasts in the policy position of scientists and their research are seen as :

(i) due to the difference between the holistic, integrative approach taken by the Garden Island scientists as the servants of management, in contrast to the highly specialised, reductionist approach to research taken by the Rottnest scientists ; and

(ii) the different policy-making and institutional contexts in which the scientists operated.

Some historical background clarifies point (i). In the 1950s and 1960s the Rottnest quokka population was used as a ready source of subjects for studies in animal physiology, and medical research into muscular dystrophy. During this time 700 animals were transferred annually to the University of Western Australia (UWA) for research purposes and the island population was encouraged to increase. However, this increase limited the natural regrowth of vegetation and frustrated the efforts of the Rottnest Island Board to reafforest the island, as the quokkas found young shrubs and trees especially palatable. Proposals by the Board in 1960 to reduce the quokka numbers, by poison, shooting and using the animals as crayfish bait, were opposed, particularly by the scientists from the WA Museum, UWA and the Department of Fisheries and Fauna. The research value of the quokka was emphasised by Dr W.D.L. Ride of the Museum, who insisted that no obstacle must challenge the medical research being carried out at the time[36]. However, while biological research influenced island environmental policy and management, it was not always made available for management purposes.

The problem of the overpopulation of quokkas re-emerged in 1989 in a debate between island staff and specialist biological scientists. The latter had retained their strong vested interest in the species, occupying positions on the advisory committees, and had dominated policy. George Seddon, a member of the Rottnest Island Foundation, and of the Environmental and Research Advisory Committee, commented on the use of Rottnest by scientists : " one might conclude that although there has been a great deal of research on the island there has been little research for the island... The quokka has long been isolated on Rottnest. The Western Australian scientific community has been isolated on its own island, and this sometimes shows "[37].

Another commentator, a former staff member (name withheld upon request), remarked that the specialist scientists on Rottnest created an exclusive

36. T. Sten, " An Appreciation of Progress in Reafforestation of Rottnest Island : 1954-1974 ", *op. cit.*

37. G. Seddon, " Vegetation Dynamics ", paper presented at the Rottnest Island Quokka Workshop, Mar. 9-10, records of the *Terrestrial Ranger*, Rottnest Island Authority, 1991.

pool of knowledge which was not always shared with the wider community. A gap existed between the science (facts) generated and management tools owing to the reluctance of the scientists to devote their resources or integrate their interests with management. The view that the scientists had a " proprietorial " attitude to Rottnest has been echoed by past and present Authority staff, committee members and the Rottnest Island Society (a community lobby group). Whereas the staff regarded the retention of scientific data as an administrative issue, the Society considered the absence of any mention of scientific work in the Authority's Annual Reports to be a lack of public accountability[38]. These perceptions point to the other major contrast.

There is a clear difference between the highly specialised, reductionist approach to research taken by the Rottnest scientists in the 1950s and the more recent holistic, ecological, integrative approach taken by the Garden Island scientists. The value of specialist studies is not denied, for example, the studies carried out on the distribution of tropical marine organisms are of world-wide interest[39]. However, the narrow approach taken by the specialists as a group towards a public island and reservation did not appear to support conservation and sustainable environmental management. These scientists were not subject to public scrutiny and operated according to their own research agenda. In contrast, the Garden Island scientists integrated research into the construction of the naval base, and into a management plan designed to maintain biological conservation, and to accommodate the multiple resources uses of Garden Island. This integration illustrates the application of holistic, ecological and geographical (Humboldtian) perspectives.

(2) The second contrast is that of the role played by the different institutions — the Rottnest Island Board/Authority and its advisory committees, and the Commonwealth Parliamentary Public Works Committee. The policy-making function of the State Government body, and the policy-review function of the federal body both provided (and continue to provide) opportunities for scientists to influence the development of public policy.

The timing of the naval base construction on Garden Island coincided with increasing public awareness of, and demand for, environmental considerations. At the global scale, the environmental movement was beginning to make an impact upon politics and policies in the western democracies, and Australia

---

38. Rottnest Island Society, " Perceived shortcomings in the management of Rottnest ", list prepared by Rottnest Island Society Secretary, Mr O. Mueller for the author, 2 July, 1995.

39. E.P. Hodgkin, L. March, G.G. Smith, " The littoral environment of Rottnest Island ", *J. Roy. Soc. W. Aust.*, 43, n° 3 (1959), 85-88 ; F.E. Wells, " Zoogeographical importance of tropical marine mollusc species at Rottnest Island, Western Australia ", *W. Aust. Nat.*, 16, n°s 2 and 3 (1985), 40-45. ; J.B. Hutchins, " Dispersal of tropical fishes to temperate seas in the southern hemisphere ", *J. Roy. Soc. W. Aust.*, 74 (1991), 79-84 ; F.E. Wells, D.I. Walker, H. Kirkman, R. Lethbridge (eds), " The Marine Flora and Fauna of Rottnest Island, Western Australia ", *Proceedings of the Fifth International Marine Biological Workshop held at Rottnest Island in Jan. 1991*, vol. 1, Perth, 1993.

was no exception. Therefore in the late 1960s and early 1970s the idea that policy provisions for major projects should include an environmental component, for which integrative studies were central, was well-received by local decision-makers. Whereas on Rottnest, the scientists became involved in the late 1940s and 1950s. This era was the start of the " modern sophisticated period of research " in Western Australia[40], which coincided with the appointment of UWA's Professor H.W. Waring and the Museum's Dr W.D.L. Ride. Increased funding from government and private benefactors and the replacement of the amateur naturalist as researcher and policy adviser by the specialist professional, created an environment for independent scientific research. In that period the links between marsupial research and nature conservation were incipient, and at least a decade would pass before links would be made to public policy[41]. Local scientific activity reflected the widely-held perspectives of the post-war era, which led to the restriction of inquiry through the direction of research towards more specialised problems[42].

CONCLUSION

The Analysis tells us something about :

(i) the islands — bio-physical : the effect on the vegetation ;

(ii) science — the narrow approach of the specialists versus the wider approach of the generalists ;

(iii) public policy — the results produced by different democratic institutions.

(1) The natural regrowth of vegetation on Rottnest following years of disturbance reflected the influence of scientists on policy. First, scientific interest combined with official protection and the human provision of summer water and food in the Settlement, encouraged the quokka population to expand beyond island carrying capacity. Second, there was no Authority requirement for, and little interest was shown in, integrative studies of the relationship between the fauna, vegetation and soils, until the 1970s[43]. In contrast, the denser vegetation on the undeveloped parts of Garden Island has largely recovered from the disturbances since colonisation, owing to the lack of interest shown in the island by the general public and researchers and the lower population of grazing tammars.

40. D.L. Serventy, " History of zoology in Western Australia ", *J. Roy. Soc. W. Aust.*, 62, n° 1-4 (1979), 33-43.

41. A.R. Main, M. Yardev, " Conservation of macropods in reserves in Western Australia ", *op. cit.* ; A.R. Main, " The study of nature - A seamless tapestry ", *op. cit.*

42. M. Bowen, *Empiricism and Geographical Thought : From Francis Bacon to Alexander von Humboldt*, Cambridge, 1981, 1.

43. D. O'Connor, C. Morris, J.N. Dunlop, L. Hart, H. Jasper, I. Pound, *Rottnest Island A National Estate Survey of Its History, Architecture and Environment. Environment*, Perth, 1977, Book 2.

(2) The cases discussed point to the wider implications of two contrasting philosophical approaches to science, and highlight a dialectic within the sciences. The rise, or re-emergence, of holism and integration, the wider view, is shown in case of Garden Island, whereas the Rottnest case shows the adherence to " traditional " reductionist, or specialist, approaches. The dialectic continues, as the management of Rottnest has changed, since the late 1980s, to take a wider, more general view, which integrates the fauna, vegetation, soils, landscape, cultural associations and land uses, rather than treating them in isolation[44]. On Garden Island it remains to be seen if the integrative perspective of the management plan is fragmented over time by specialist interpretations of the plan.

These local contrasts reflect the wider historical dialectic within the sciences between the philosophy of dualistic, positivist empiricism and holistic, " social empiricism "[45].

(3) Although the contrasting influences of scientists on public policy occurred within different political contexts, there is also an on-going dialectic, in regard to openness to public scrutiny, within the administrative institutions for both Rottnest and Garden Island. On one hand, the inquiries made by the PWC on the Garden Island works meant that the opinions and concerns of experts outside the public service, community groups and State and local governments were not only heard, but made a contribution to policy. However, since the establishment of the GIEAC, it has produced only one public report of its activities and is not open to public review[46]. While the advisory committee system of the Rottnest Island Authority does not allow for public scrutiny of decisions, there has been, since 1987, a legislative requirement for a Rottnest management plan to be open to public input. This directive may not make the policy decisions of the advisory committees and the board more transparent, but it does provide an opportunity for public review.

In conclusion, the cases of Rottnest and Garden Island have illustrated a brief segment in the history of science in Western Australia, but one reflective of a longer and broader history. Over the last fifty years there has been a change in the relationship of scientists with policy and management — from purely reductionist science, detached from other aspects of island policy and management, to more holistic science, integrated with management. The Garden Island case shows how the practical applications of the wider approach in policy making, and the management of public reserves, contributes to both knowledge and the community. When scientists of different disciplines work

44. Rottnest Island Management Planning Group, 1985 ; Government of Western Australia, 1995.

45. See M. Bowen, *Empiricism and Geographical Thought : From Francis Bacon to Alexander von Humboldt, op. cit.*

46. Garden Island Environmental Advisory Committee, *Garden Island : Environmental Advisory Committee Report of Activities 1995-1996*, Rockingham, 1997.

together, rather than apart from each other, and work with policy makers and managers, this integration takes science out of isolation, and enhances its value by broadening its scope and influence. Therefore, the particular combination of political institutions with integrative, science regarding the management of Garden Island is seen to be superior to the Rottnest system.

Nouvelles du Brésil : l'Institut historique de Paris et le projet de l'écriture de l'histoire du Brésil de l'*Instituto Histórico e Geográfico Brasileiro*

Alda HEIZER

> " Lorsque le roi D. João VI a changé vers le Rio de Janeiro,
> la place de son Empire, le Brésil s'ouvrit aux étrangers ".
>
> (Auguste de Saint-Hilaire, 1816)

Pendant le XIXᵉ siècle une quantité considérable de voyageurs, artistes et savants ont enregistré dans leurs journaux, rapports, tableaux et dessins leur vision du Brésil. Ce sont des registres qui révèlent comment le Brésil a été vu et pensé par ceux qui ont participé à des voyages exploratoires : un pays exubérant, exotique, mais sans tradition et sans histoire.

L'IHGB a été créé en 1838 au sein d'un projet modernisateur d'un Etat en train de se consolider. Ce projet se présentait dans les différentes propositions de ceux qui se donnaient comme but la construction d'un Empire qui puisse être reconnu par les " nations civilisées " : la France et l'Angleterre.

Depuis l'arrivée de D. João VI et de la cour portugaise à Rio de Janeiro, en 1808, jusqu'à la moitié du XIXᵉ siècle ont été créées diverses institutions et leurs conceptions et fonctionnement étaient liés à la couronne[1]. Il s'agissait d'un Etat promoteur et protecteur de la culture.

L'IHGB n'a pas été une exception. Cette institution a été créée à partir de la *Sociedade Auxiliadora da Indústria Nacional*[2]. L'IHGB, organisé selon le modèle des académies littéraires de la fin du XVIIᵉ et XVIIIᵉ siècles, a suivi la

---

1. Academia de Belas Artes (1810) ; Biblioteca Real (1810) ; Museu Real (1818) ; Imperial Observatório (1827) ; Arquivo Nacional (1838) par exemple.

2. La SAIN (*Sociedade Auxiliadora da Indústria Nacional*), créé en 1827 avait pour objectif de " vulgariser des connaissances utiles à l'agriculture et aux autres activités productives de la Nation ". L. Werneck da Silva, *Assim é o que me parece. A Sociedade Auxiliadora da Indústria Nacional 1827-1904 na formação social brasileira*, Rio de Janeiro, 1978.

tradition française de l'Institut Historique de Paris, créé en 1834. Cet Institut français a été idéalisé par Eugène Gary Monglave et il avait pour objectif d'organiser " une société d'érudition qui s'occupait de la recherche à caractère historique "[3]. Il comptait sur l'appui du ministre Guizot[4].

Des intellectuels brésiliens ont participé à la fondation de l'organisation initiale de l'Institut français. Manuel de Araújo Porto Alegre, Domingos José Gonçalves et Francisco Sales Torres Homem ont été parmi les fondateurs de l'Institut Historique de Paris. En outre, cet institut français, jusqu'à la fin du XIXe siècle, a soutenu une large correspondance avec l'institut brésilien. L'IHGB était composé d'un groupe social appelé par un mémorialiste de l'Empire " de la bonne société "[5]. Il s'agissait d'une élite intellectuelle composée par des hommes liés à l'Etat Impérial.

Fondateurs[6] de l'IHGB :

> Alexandre Maria Mariz Sarmento
> Aureliano de Sousa e Oliveira Coutinho (député)
> Bento da Silva Lisboa
> Caetano Maria Lopes Gama (député)
> Cândido José de Araújo Vianna (député)
> Francisco Cordeiro da Silva Torres Alvim
> Francisco Gê de Acaiaba Montezuma (député)
> Januário da Cunha Barbosa
> Joaquim Francisco Vianna (député)
> José Antônio Lisboa
> José Antonio da Silva Maia (député)
> José Clemente Pereira (député)
> José Feliciano Fernandes Pinheiro (sénateur)
> Raymundo José da Cunha Mattos
> Rodrigo de Sousa Silva Pontes

La tâche des membres de l'IHGB consistait dans l'écriture d'une histoire du Brésil qui était " un déplissage aux tropiques d'une civilisation blanche et européenne "[7].

L'un des objectifs de l'IHGB était de stimuler et de promouvoir des voyages exploratoires ayant pour objet de recueillir des données qui puissent aider à

3. M.A. de Oliveira Faria a publié les résultats de ses recherches sur le séjour des romantiques brésiliens à Paris, 1833-1836, dans la *Revista do IHGB*, sous le titre " Os brasileiros no Instituto Histórico de Paris ", vol. 266 (jan.-mars 1965), 68-148.

4. Voir L. Thiers, " Guizot et les Institutions de Mémoire ", dans P. Nora (éd.), *Les Lieux de Mémoire*, Paris, 1986.

5. F. de Paula Ferreira de Resende a caractérisé un groupe social privilégié de la société esclavagiste et coloniale brésilienne.

6. L.M. Paschoal Guimarães, " Debaixo da Imediata Proteção de sua Majestade Imperial ", *R. IHGB, RJ*, 156 (388) (Jul/Set. 1995), 459-613.

7. M.L. Salgado Guimarães, " Nação e Civilização nos Trópicos : o ihgb e o projeto de uma história nacional ", *Revista de Estudos Históricos*, n° 1 (1988), 5-28.

l'écriture de l'histoire du pays et de chercher dans les archives étrangères de *Nouvelles du Brésil*, afin d'utiliser le titre d'une section du *Journal de L'Institut Historique de Paris.*

Dans le projet de la revue trimestrielle de l'IHGB il y avait trois thèmes fréquents : la question indigène, les voyages et les explorations scientifiques ; les discussions sur l'histoire régionale[8].

Les voyages exploratoires au Brésil étaient demandés par les gouvernements brésiliens et étrangers et ils recouvraient une mission civilisatrice. Les matériaux pris par les étrangers deviennent autant d'images du Brésil qui restent distribuées dans diverses institutions européennes. Ces vestiges de l'histoire du Brésil pouvaient être trouvés dans le Musée d'Antiquités Américaines de Copenhague, par exemple. Ce Musée était lié à la Société Royale des Antiquaires du Nord, comme l'*Academia Real de Lisboa* et comme la Société de Géographie de Paris, et gardait des rapports savants avec l'IHGB.

Les liens existants entre l'IHGB et l'Institut Historique de Paris peuvent être constatés à travers l'échange de documents entre les deux instituts. En examinant ces documents (discours et rapports savants), on peut repérer une même vision de l'histoire.

" A fundação do IHGB é uma grande e feliz idéia... Começar pela Geografia e História é começar bem, é lançar uma vista sobre o passado, para obter esclarecimento, que sirvam de illuminar todos os momentos de tempo presente... "[9].

" Por fim devo ainda ajuntar uma observação sobre a posição do historiador do Brasil para com sua pátria. A história é uma mestra, não somente do futuro, como também do presente... "[10].

Dans les discours de ceux qui étaient occupés de penser le Brésil le rôle de l'histoire était clair dans la construction de l'Etat Impérial.

Après la fin de la monarchie de Louis Philippe, l'Institut français continue ses rapports avec l'IHGB et le discours qui mettait en valeur la double fonction de l'histoire se renforçait : l'histoire est maîtresse de la vie et, au même temps, un instrument pour les hommes de la " bonne société " se constituer en " classe seigneuriale ".

On devrait remarquer comment la tradition française héritée de l'Institut Historique des Paris se trouve dans le discours de ceux qui ont eu la tâche d'écrire l'histoire du Brésil — l'IHGB — et les rapports de ce projet intellectuel avec la construction d'un Empire aux tropiques.

---

8. M.L. Salgado Guimarães, " Nação e Civilização nos Trópicos : o ihgb e o projeto de uma história nacional ", *op. cit.*, 20.

9. E. Gary Monglave, Lettre écrite de Paris au secrétaire de l'IHGB, Januário da Cunha Barbosa, au 22/10/1839. *Revista do IHGB*, vol. I, 1839.

10. K.F.P. von Martius, " Como se deve escrever a história do Brasil ", *Revista do IHGB*, vol. VI, 1845.

L'IHGB centre intellectuel autorisé par le gouvernement impérial, gardait et vulgarisait les registres des voyageurs étrangers et, ainsi, contribuait pour que l'Empire puisse intégrer l'ensemble des " pays civilisés " bien que cet Empire fût obligé de supporter l'existence du noir et de l'indigène. Ces deux groupes, selon le discours des voyageurs européens et des intellectuels des deux instituts étaient destitués de toute civilisation.

L'IHGB, le premier espace institutionnel de la production historiographique au Brésil, vu les caractéristiques de sa formation culturelle, souffre l'influence d'une " société d'érudition " : l'Institut Historique de Paris. Le modèle français a gouverné l'organisation et la production culturelle de l'IHGB.

Les premiers contacts avec ce thème ont permis de reconnaître les intellectuels des deux instituts qui ont participé à la vie politique et des institutions liées à l'Etat. L'Empereur Pedro II, par exemple, a été " protecteur " des deux institutions.

Entre 1834 et 1836, dans les registres du *Journal de l'Institut Historique* figurent 48 membres brésiliens qui ont participé aux différentes commissions[11]. Il s'agit d'un groupe hétérogène d'intellectuels — poètes, diplomates, écrivains — et, pour la plupart, des hommes liés à l'Etat. " Criavase um lugar onde os intelectuais forjariam a nacionalidade brasileira através de uma perspectiva histórica marcada pela idéia de progresso linear : do estado de barbárie a nação brasileira passava a um estado de civilização e progresso. Esta perspectiva, presente no século XIX, quando pensar a história é uma constante, vai caracterizar o debate historiográfico deste momento[12]. "

La " matière première " de ceux qui se donnaient la tâche d'écrire l'histoire du Brésil incorporait les rapports de voyageurs français qui ont été au Brésil pendant la première moitié du XIXe siècle. Ces rapports se trouvent aux archives et musées européens et présentent une vision du Brésil homogénéisée, pour la plupart.

Les voyageurs français apportaient, outre les images d'un Brésil exotique et exubérant, un point de vue favorable au projet de construction d'un empire aux tropiques. Jean Baptiste Debret, par exemple, venu au Brésil invité par D. João VI dans la mission artistique française en 1816, a enregistré le quotidien de la colonie et il a présenté à l'Institut Historique de Paris une vision favorable à la présence française dans la ville de Rio de Janeiro. Auguste de Saint-Hilaire a été au Brésil en 1820 et a enregistré dans ses livres des descriptions importantes d'un regard européen et civilisateur sur la colonie.

L'IHGB a stimulé l'organisation et la réalisation de voyages exploratoires et la publication des résultats de ces voyages et il a contribué au projet d'intégra-

---

11. Voir M. Alice de Oliveira Faria, " Os Brasileiros no Instituto Histórico de Paris ", *Revista do Instituto Histórico*, 266 (1965), 118.

12. Voir A. Lùcia Heizer, *Uma Casa Exemplar. Pedagogia, Memória e Identidade do Museu Imperial de Petrópolis*, Dissertação de Mestrado PUC/RJ, 30.

tion physique du pays en définissant l'espace géographique brésilien et la place de chacun dans ce projet de nation.

Parmi les préoccupations des dirigeants de l'Etat, on peut reconnaître la démarcation du territoire. Les conflits de la régence, les mécontentements de la présence portugaise au Brésil, l'insertion des noirs et des indiens menaçaient, selon cette perspective, l'ordre interne du pays. D'autre côté, il y avait la menace externe qui se traduisait par les récentes indépendances des républiques latino-américaines. Ainsi, la démarcation géographique de la nation était urgente et l'installation d'un Etat puissant capable de maintenir l'ordre.

Donc, je crois possible de reconnaître la vision et le discours civilisateur des registres des voyageurs français qui ont été au Brésil entre 1810 et 1870 et le rôle de ces registres dans l'histoire du Brésil proposée par l'IHGB.

Sur les géographes de l'entre-deux-guerres. Cartographie " appliquée ", nationalismes, planification

Marie-Claire Robic

L'expression d'" application " des sciences de l'homme suggère qu'il s'établit une relation immédiate entre une science déjà là et un champ de pratique qui serait extérieur à la discipline. Ce champ importerait directement l'appareillage conceptuel et méthodologique de la science de référence, ou bien, en un mouvement inverse qui suppose aussi l'absence de médiations, dicterait ses propres intérêts et les imprimerait sans traduction dans la science. L'analyse de l'usage de la cartographie par les géographes de l'entre-deux-guerres suggère qu'un système d'interactions plus complexe se noue entre les pratiques cognitives d'une discipline et les expertises auxquelles elle prétend contribuer. On évoquera surtout les activités des géographes mêlés aux Congrès internationaux de géographie tenus sous l'égide de l'Union géographique internationale (UGI) pendant les années trente, en élargissant la période d'observation à plusieurs décennies, jusqu'au-delà des années cinquante où la cartographie s'épanouit et où s'est produite une réorientation significative de la géographie vers la perspective cognitive et pragmatique de l'" organisation de l'espace géographique ". On essaiera par cette étude de cas de poser les éléments d'une argumentation qui montre que les transformations sensibles dans les rapports qu'entretient une discipline comme la géographie avec la pratique doivent se concevoir dans un jeu entre la dynamique de savoirs disciplinaires et la formulation (dynamique elle aussi), de projets d'intervention collective dans la sphère politique et sociale.

De l'inventaire du monde à l'engouement pour la cartographie " appliquée "

L'étude des " voeux " ou " résolutions " que les membres des Congrès internationaux de géographie ont régulièrement émis des années 1870 à 1950 révèle une transformation générale de leur attitude vis-à-vis de la Cité[1]. En résumé, entre la fin du XIXᵉ siècle et les années trente, ils passent d'une position générale de " conseil aux gouvernants ", voire de *lobbying* en direction des grandes puissances, à une organisation autocentrée, une polarisation sur la " Cité scientifique " des géographes. La Grande Guerre marque la coupure.

Qu'en est-il de leur usage de la carte, cet outil souvent jugé spécifique aux géographes ? Un profond changement s'observe en parallèle. Le premier moment, celui des congrès du XIXᵉ siècle gérés par les sociétés de géographie, s'accompagnait d'un investissement scientifique centré sur l'inventaire cartographique du monde. Les congressistes avaient notamment comme objectif la réalisation d'une carte standardisée du globe, à l'échelle du millionième. Ce monde, que l'on considérait comme désormais connu dans ses formes principales, était aussi entièrement approprié par les puissances occidentales. C'était un monde " fini ", à représenter de manière exhaustive et homogène. Les cartes reines du géographe de l'époque étaient alors d'une part la carte dite " topographique ", une carte à grande échelle représentant le relief et l'hydrographie ainsi que les marques de l'occupation humaine, et, d'autre part, la carte à petite échelle. Dans les deux cas, la valeur de la cartographie reposait sur des moyens considérables relevant des États-majors ou de grands services d'État. En relevaient les topographes, les ingénieurs-géographes, les géodésistes, etc., spécialistes civils ou militaires de " sciences géographiques ", qui côtoyaient aux Congrès des géographes devenus au XXᵉ siècle, majoritairement, des universitaires.

Ces professeurs de géographie partageaient l'intérêt des cartographes-géographes pour les deux types de cartes. Non seulement leur engagement dans leur réalisation était extrêmement fort, mais, comme on l'a montré à propos des géographes français du début du siècle, ils se montraient très critiques à l'égard des procédés cartographiques utilisés par les statisticiens depuis le milieu du XIXᵉ siècle[2] : ils étaient extrêmement réticents à l'emploi de cette cartographie que l'on appelle depuis les années 1960 la cartographie " thématique "[3]. Or le second moment de l'évolution d'ensemble des géographes (dans leur association internationale du moins), caractérisé par leur retrait dans la communauté scientifique, se marque par un véritable engouement pour cette cartographie thématique que pendant l'entre-deux-guerres l'on taxe de " spéciale " ou d'" appliquée ". Que recouvre ce recours massif à la carte dans un moment où les géographes tendent à se centrer sur leur spécificité disciplinaire ?

---

1. M.-C. Robic, A.-M. Briend, M. Rössler (eds), *Géographes face au monde. L'Union géographique internationale et les Congrès internationaux de géographie*, Paris, Montréal, 1996.

2. G. Palsky, *Des chiffres et des cartes. Naissance et développement de la cartographie quantitative française au XIXᵉ siècle*, Paris, 1996.

3. L'expression a été créée par les auteurs allemands au tout début des années cinquante.

La carte comme outil de recherche pure et d'expertise du territoire national

L'engouement des années trente pour la cartographie thématique présente des dimensions contrastées.

Il présente de multiples facettes. Il est d'abord très général, puisqu'il se produit dans toutes les branches de la discipline — de telle sorte que la diffusion des innovations n'est pas le fait des cartographes-géographes (restés au demeurant très classiques) — et puisqu'il concerne toutes les " écoles " nationales présentes. En second lieu, il s'agit moins d'invention graphique que d'usage de codes sémiologiques définis antérieurement par les statisticiens. Mais la pratique des géographes étend considérablement la gamme des phénomènes représentés dans les domaines de la géographie humaine, économique et sociale. En outre, les géographes manient une cartographie beaucoup plus sophistiquée qu'on ne l'imaginerait aujourd'hui, ce qui souligne l'intérêt profond qu'ils portent à ce mode d'expression et leur maîtrise des systèmes cartographique et statistique : l'usage des valeurs centrales et de la dispersion statistique, des doubles gammes de valeurs, des isolignes, de la statistique spatiale, etc., est fréquent. Enfin, une différenciation interne à la communauté scientifique transparaît sous cet engouement, révélant une forte segmentation nationale : dans tel pays, comme en Pologne, l'intérêt se porte sur la mise en valeur des disparités du développement national, grâce à l'emploi de la statistique différentielle ; là, en URSS, l'accent est mis sur le déplacement du centre de gravité du territoire ; la Grande-Bretagne excelle dans la cartographie d'inventaire ; l'" école française " reste fidèle à la carte topographique et se contente d'une cartographie quantitative élémentaire.

Comme le montrent les discussions auxquelles est mêlée la référence à la carte, cet engouement manifeste que les géographes tendent à établir un nouveau rapport à la cartographie. Celle-ci devient une méthode intégrée à la démarche géographique et qui devrait s'avérer efficace à trois points de vue. Elle est conçue comme l'instrument de prédilection qui pourrait construire une communauté scientifique en la dotant d'un outil commun de recherche et d'évaluation des résultats. La carte est aussi le moyen par excellence de contrôle et/ou de test des théories (ceci valant à l'époque pour la géographie physique essentiellement). Partout où la démarche inductive est préférée — et c'est ce qui domine —, la carte devient la base d'enquêtes comparatives, telle l'enquête internationale sur l'habitat rural, dont le projet oscille entre l'établissement de types régionaux et la recherche de corrélations significatives entre milieu et habitat.

A côté de cette contribution de l'outil cartographique à la recherche pure, trois types de considérations construisent le projet d'une cartographie proprement géographique qui serait l'instrument d'une expertise territoriale.

Plusieurs interventions, traduites en résolutions de congrès, estiment que la carte et la graphique devraient devenir le médium privilégié d'une communication avec l'extérieur de la géographie. Celle-ci vise moins le grand public qui s'ouvre à une consommation géographique induite par le tourisme populaire (c'est l'époque des " congés payés "...), que de nouveaux interlocuteurs publics et privés. Ainsi les services gouvernementaux voués à la planification territoriale, qui se multiplient au moment de la Grande Crise, et les organismes économiques, telles les entreprises de transport, sont les " cibles " des géographes[4].

La carte thématique devient le moyen d'observation des différences spatiales, des inégalités affectant les territoires nationaux. Son utilité s'affirme d'abord pour appréhender globalement le degré d'homogénéité d'un pays. En outre elle met sur la voie d'une gestion corrective, ou en tout cas d'une rationalisation de l'espace.

La Pologne est exemplaire de la volonté transformatrice qui accompagne la description de disparités internes héritées d'une histoire territoriale conflictuelle. D'ailleurs des observateurs soulignent la portée politique de l'investissement cartographique dans les nouveaux pays de l'Europe centrale ; d'autres remarquent que le contexte explique la multiplication, en Europe, d'atlas nationaux où s'impose une cartographie " non conventionnelle "[5].

Les géographes les plus proches des milieux où une planification nationale a été instituée franchissent le pas de l'offre de service. Ils affirment que la géographie, en se servant de l'outil cartographique, est la discipline-clé d'une planification qui se cherche : une planification *territoriale* et non pas sectorielle, où la cohérence de l'action publique résiderait dans la base spatiale que la carte permet de représenter d'un seul coup d'oeil.

Dans la tension nationaliste, un problématique " marché " de l'aménagement ?

En fait, derrière l'apparence d'une Cité scientifique auto-organisée, la collectivité des géographes des années trente est en proie à des tensions politiques considérables. L'une des plus fortes repose sur la rivalité franco-allemande,

---

4. *Cf.* un voeu du Congrès de 1938 : " La section de géographie économique (…) émet le voeu que la détermination qualitative et quantitative des divers modes de transport soit facilitée par l'accumulation de données statistiques élaborées le plus possible de manière uniforme et qu'une collection de cartes et de cartogrammes de géographie du trafic se rapportant au plus grand nombre possible de pays soit présentée au prochain congrès, afin de démontrer dans quelle mesure le spécialiste de géographie économique pourra fournir un travail préparatoire efficace à la solution scientifique du problème de coordination du trafic ".

5. *Cf.* une recension d'atlas : " Intensified national feeling, movement towards social and economic planning, and rising standarts of education have all created a demand for special atlases of particular nations and regions and for the inclusion of specialized maps in the general atlases ". (Anon., " Six representative European atlases ", *The Geographical Review*, 1937, 161-163, citation p. 161).

vive dès avant 1914 et réactivée par la Grande Guerre. Elle s'est traduite par l'exclusion des géographes allemands de l'Union géographique internationale. A partir de 1934, année de leur entrée dans l'organisation internationale sous la pression de Hitler, le conflit a sous-tendu toutes les prises de parole.

Mais, plus généralement, les débats des années trente ont eu pour toile de fond une forte tension nationaliste, que l'affirmation nationale repose sur un simple affichage de l'intérêt propre ou sur une volonté d'hégémonie. En rupture avec les représentations universalistes des rencontres d'avant-guerre, les logos des Congrès internationaux manifestent graphiquement cette tension : en 1934 à Varsovie et en 1938 à Amsterdam, la carte du pays d'accueil se substitue au globe ou à la mappemonde ornant les logos antérieurs...

Ces colloques des années trente sont les premières rencontres à vocation scientifique entre des géographes qui, dans leurs pays respectifs, ont l'expérience d'une pratique de géographie " appliquée " inscrite dans l'aménagement spatial, c'est-à-dire dans un champ d'intervention nouveau pour la discipline[6]. Mais ils y participent diversement. Soit ils contribuent à des entreprises de planification conçues comme une simple mais nécessaire extension de l'urbanisme (celui-ci est né dans les premières décennies du siècle). Soit ils sont eux-mêmes les promoteurs d'interventions nouvelles (tel le *Land Utilization Survey* réalisé à l'initiative de Dudley Stamp avec l'appui de la *London School of Economics*). Soit ce sont des praticiens engagés dans l'urgence dans des opérations officielles destinées à remédier à la Crise, tel le *New Deal*. Soit enfin l'entreprise répond à une visée totalitaire, comme dans le régime nazi, où les géographes, avec d'autres scientifiques, ont été mobilisés dans des services centralisés dédiés à l'étude de l'espace (*Raumforschung*) et à l'organisation de l'espace (*Raumordnung*)[7]. Les vocables nouveaux de *Regional* ou *National planning*, *National Survey, Landes-* ou *Raumplanung, Bonificazione...*, inscrivent ces entreprises collectives dans l'action politique. Les géographes qui s'y impliquent participent souvent d'une visée de rationalisation du social largement partagée dans le monde occidental depuis le début du XXe siècle, et qui s'est traduite notamment par la montée de l'idéologie technocratique.

Les communications discutées aux Congrès montrent bien l'intensité et l'ampleur de l'intérêt suscité par l'établissement d'une évaluation des divers territoires nationaux.

L'attention portée à ses transformations récentes, à ses disparités, à la planification qui s'y applique, révèle bien — avec d'autres marqueurs tels les questions admises au programme officiel, les discussions et les vœux —, l'acuité de la question.

6. Voir V. Berdoulay, H. van Ginkel (eds), *Geography and professional practice*, Utrecht, 1996.

7. M. Rössler, " Applied geography and area research in Nazi society : central place theory and planning, 1933 to 1945 ", *Environment and planning*, D : *Society and space*, 7 (1989), 419-431.

Les analyses de la crise soulignent que les localisations modernes relèvent d'un monde en transformation, à l'échelle de la planète comme à celle des pays développés : la dimension mondiale de la concurrence est admise, et l'urbanisation, la mobilité liée à l'automobile, la croissance des services, paraissent les facteurs d'un changement spatial irréversible dont les géographes peuvent dégager les logiques.

Les débats s'ouvrent donc souvent sur des considérations pragmatiques relatives à la manière dont la géographie, au nom de son savoir propre, peut contribuer à une action volontaire sur le territoire (au " re-planning " disent parfois les Anglais).

Mais tout se passe comme si les expériences étaient trop distinctes de pays à pays, comme si elles étaient trop fragmentaires pour l'ensemble des géographes, comme si le destin de chaque pays était trop particulier, comme si enfin les conflits politiques étaient trop présents, pour qu'une formulation claire et unanime des enjeux nouveaux soit possible. Quelques cas sont révélateurs de cet état d'*émergence* d'un champ de pratiques géographiques potentielles qui ne s'affirmeront qu'après la Seconde Guerre mondiale.

Ainsi de la carte et de sa fonction. Il n'y a pas de commune mesure entre les timides esquisses d'une iconographie scientifique nouvelle et l'agressivité graphique que savent déployer au même moment, hors de la communauté scientifique, des géographes géopoliticiens plus soucieux de propagande que de communication et de recherche. Si l'on prend le cas français, c'est seulement après 1950 que se développera une réflexion sémiologique sur la carte, attachée à l'analyse de l'information et à l'efficacité du message visuel[8] ; la cartographie thématique se diffusera largement dans les atlas régionaux, tandis que l'aménagement du territoire sera l'un des meilleurs utilisateurs d'une cartographie nouvelle, comme le montrent les publications statistiques de la DATAR[9].

Ainsi du destin de débats épistémologiques et de théories présentées aux Congrès des années trente. C'est là que s'amorcent des discussions conceptuelles sur la région, sur le paysage, sur la ville, qui ne s'épanouiront aussi qu'après la guerre. Elles auront alors une connotation aménagiste (pour la ville et la région notamment). Elles concerneront les géographes et de nouveaux venus, les économistes spatiaux qui créent dans le même temps la Regional Science.

Les deux notions de région et de paysage, mêlées dans le terme allemand de Landschaft, entraient dans les opérations d'aménagement engagées pendant l'entre-deux-guerres. Les perspectives d'une application normative des régularités spatiales recherchées par les théoriciens des localisations n'étaient pas

---

8. *Cf.* les travaux de J. Bertin, *Diagrammes, réseaux, cartographie. Sémiologie graphique*, Paris, 1967.

9. Délégation à l'aménagement du territoire et à l'action régionale, créée en 1963.

non plus étrangères aux promoteurs de ces recherches non-conformistes, tel un W. Christaller[10] : mais l'étrangeté de leur pratique heuristique dans une discipline dominée par une épistémologie inductive explique leur très faible écho dans les Congrès des années trente. Le succès viendra à partir des années cinquante, et d'abord aux États-Unis. Ces théories constitueront la base de modèles de planification durant les années soixante. Elles triompheront également en patronnant la *new geography* qui s'est construite sur l'opposition entre la démarche idiographique attribuée à la géographie " classique " et une démarche nomothétique à promouvoir sur le modèle scientifique standard, et sur la définition du nouvel objet de la discipline : l'" espace géographique ".

En somme, dans le contexte tendu des années trente, les échanges internationaux sur des expériences segmentées, particularistes, ne pouvaient provoquer une conscience forte des enjeux notionnels, épistémologiques, pratiques, ouverts par l'émergence d'un potentiel de géographie appliquée au développement national. Encore moins possible sans doute était la construction collective d'un nouveau champ de l'action (l'organisation de l'espace) réclamant des concepts et des outils nouveaux, mais aussi des institutions et des financements... Ce n'est qu'après la Seconde Guerre mondiale que se sont trouvés réunis les facteurs susceptibles de constituer des nouveaux " marchés " de la géographie. En témoigne le développement, sur le plan international comme dans tous les pays, de commissions de " géographie appliquée " qui ont tenté de susciter des pratiques d'aménagement spatial, en même temps que de s'arrimer au nouveau champ de compétences ouvert par des politiques d'aménagement du territoire. La cartographie s'est trouvée promue dans ce mouvement qui a d'une part banalisé la représentation dite thématique et d'autre part rénové la représentation géographique par l'invention de nouvelles formes de modélisations graphiques liées à la sémiologie et aux recherches de la " nouvelle géographie ". Au total, ce survol, sur une durée d'un bon demi-siècle, des changements intervenus dans les savoir-faire cartographiques et dans les ambitions des géographes en matière d'application embrasse plusieurs niveaux : celui des fondements épistémologiques de la discipline (l'évolution pouvant aller jusqu'à la révolution paradigmatique engagée dans les années cinquante) ; celui des compétences, plus proches du " métier " que de la discipline scientifique ; celui des pratiques sociales, qui ont ouvert l'éventail de la " profession géographe " hors de la fonction enseignante. La transformation des uns et des autres a pu conduire à des concurrences avec des disciplines et des branches professionnelles nouvelles ou déjà affirmées. Le concept d'organisation de l'espace qui a accompagné le changement traduit bien par son ambivalence la rencontre entre des dynamiques sociales et cognitives.

---

10. Auteur d'une théorie des villes admise seulement après la guerre (W. Christaller, *Die zentralen Örte im Süddeutschland...*, Iena, 1933).

CONTRIBUTORS

Svetlana AKHUNDOVA
Geology Institute
Baku (Azerbaijan)

Carlos ALMAÇA
Museu Bocage
Lisboa (Portugal)

Manfred BÜTTNER
Ruhr Universität Bochum
Bochum (Germany)

Michael A. CREMO
Bhaktivedanta Institute
Los Angeles, CA (USA)

Steven L. DRIEVER
University of Missouri-Kansas City
Kansas City, MO (USA)

Andrés GALERA
Centro Estudios Historicos
Madrid (Spain)

Rebeca de GORTARI RABIELA
UNAM
Mexico (Mexico)

Martin GUNTAU
Ernst-Alban-Gesellschaft
Rostock (Germany)

Alda HEIZER
Museu de Astronomia
Rio de Janeiro (Brazil)

Marion HERCOCK
University of Western Australia
Nedlands, Perth (Western Australia)

Grigoriy KOSTINSKIY
Russian Academy of Science
Moscow (Russia)

Goulven LAURENT
Ministère de l'Education Nationale
Paris (France)

Iaroslav MATVIICHINE
Ukrainian Academy of Science
Kiev (Ukraine)

Alexei POSTNIKOV
Academy of Sciences
Moscow (Russia)

Marie-Claire ROBIC
Université de Paris I, CNRS
Paris (France)

María Josefa SANTOS
UNAM
Mexico (Mexico)

Ute WARDENGA
Institute of Regional Geography
Leipzig (Germany)

Tamara A. Zolotovitskaya
Geology Institute
Baku (Azerbaijan)